Radical Reactions in Aqueous Media

RSC Green Chemistry

Series Editor:
James H Clark, *Department of Chemistry, University of York, York, UK*
George A Kraus, *Department of Chemistry, Iowa State University, Iowa, USA*

Titles in the Series:

How to obtain future titles on publication:
A standing order plan is available for this series. A standing order will bring delivery of each new volume immediately on publication.

For further information please contact:
Sales and Customer Care, Royal Society of Chemistry,
Thomas Graham House, Science Park, Milton Road, Cambridge,
CB4 0WF, UK
Telephone: +44 (0)1223 432360, Fax: +44 (0)1223 420247, Email: books@rsc.org
Visit our website at http://www.rsc.org/Shop/Books/

Radical Reactions in Aqueous Media

V. Tamara Perchyonok

School of Chemistry, Monash Universiy, Melbourne, Australia

RSCPublishing

RSC Green Chemistry No. 6

ISBN: 978-1-84973-000-6
ISSN: 1757-7039

A catalogue record for this book is available from the British Library

Published by The Royal Society of Chemistry,
Thomas Graham House, Science Park, Milton Road,
Cambridge CB4 0WF, UK

Registered Charity Number 207890

For further information see our web site at www.rsc.org

Preface

Life, the most complex form of organic compounds on Earth, requires the construction of chemical bonds in an aqueous environment. Following Nature's lead, the challenge for today's chemist is to move away from highly volatile and environmentally harmful organic solvents and towards friendly and biologically compatible media. The obvious solvent of choice is water, due to its abundance, cost-effectiveness and biological compatibility. The potential usefulness of free radical reactions in water is demonstrated by the ever-increasing number of studies over the last 20 years.

Free radicals are ubiquitous, reactive chemical entities. Free radical reactions are an important class of synthetic reactions that have been traditionally performed in organic solvents. In recent years, the number of reports of free radical reactions that use water has increased. Radical reactions are one of the most useful methods for organic reactions in water, because most of the organic radical species are stable in water, and they do not react with water. In addition, by harnessing free radical reactivity within the laboratory, biological processes can be studied and controlled, leading in turn to the prevention of disease and the development of new treatments for disease states mediated by free radicals.

Although there have been several excellent reviews on carbon–carbon bond formation and reactions of carbon–hydrogen bonds in water, this book addresses homolytic bond formation in aqueous media *via* radical reactions including chain and non-chain reactions with a particular focus on carbon–hydrogen and carbon–carbon bond formation. The book is aimed to fill a gap between undergraduate chemistry and the application aspects of reliable and user-friendly free radical chemistry suitable for use by workers from undergraduates all the way to industrial chemists and academic researchers. The emphasis of this type of book lies in combining the extensive knowledge and scope of free radical chemistry in general with green free radical chemistry with the latest innovations and creative applications of bond formation already reported in recent years and in creating a 'knowledge pool' for chemistry yet to be unearthed. The sky is the limit for the applications and development in the area.

V. Tamara Perchyonok

RSC Green Chemistry No. 6
Radical Reactions in Aqueous Media
By V. Tamara Perchyonok
© V. Tamara Perchyonok 2010
Published by the Royal Society of Chemistry, www.rsc.org

Contents

RSC Green Chemistry No. 6
Radical Reactions in Aqueous Media
By V. Tamara Perchyonok
© V. Tamara Perchyonok 2010
Published by the Royal Society of Chemistry, www.rsc.org

Acknowledgements

I gratefully acknowledge the support of my parents, Faina and Lazar, and family throughout the journey and for being an essential part of it, for letting me follow my dream, for being a constant source of inspiration, encouragement and motivation. I am also indebted to enthusiastic colleagues, past and present mentors and research associates who continuously stimulated my interest in this area of science and who are too numerous to mention. I have also been privileged to meet some exceptional individuals, who have encouraged me to see outside the square through the eyes of music (Ian J. Southwood) and language (Lorenzo Grilli).

Special thanks are due to Dr Merlin Fox and Dr Anne E. Johnson for their generous assistance with this project.

CHAPTER 1

Free Radical Chemistry and Green Chemistry: The Historical Perspective

Green chemistry means environmentally friendly organic synthesis.[1] The essential aims are to reduce the amounts of dangerous, toxic starting materials and by-products (waste disposal) and to reduce damage to the natural environment. Most processes that involve the use of chemicals have the potential to cause a negative impact on the environment. It is therefore essential that the risks involved be eliminated or at least reduced to an acceptable level. Traditionally, the risks posed by chemical processes have been minimized by limiting exposure by controlling so-called circumstantial factors, such as the use, handling, treatment and disposal of chemicals. The existing legislative and regulatory framework that governs these processes focuses almost exclusively on this issue. By contrast, green chemistry[2] seeks to minimize risks by minimizing hazards. It thereby shifts control from circumstantial to intrinsic factors, such as the design or selection of chemicals with reduced toxicity and of reaction pathway that eliminate by-products or ensure that they are benign. Such design reduces the ability to manifest hazards (and therefore risks), providing inherent safety from accidents or acts of terrorism.

The most widely accepted definition of green chemistry is 'the design, development and implementation of chemical processes and products to reduce or eliminate substances hazardous to human health and the environment'. This definition has been expanded into 12 'Principles of Green Chemistry':[1]

- *Prevent waste*: Design chemical syntheses to prevent waste, leaving no waste to be treated or cleaned up.
- *Design safer chemicals and products*: Design chemical products to be fully effective, yet have little or no toxicity.

RSC Green Chemistry No. 6
Radical Reactions in Aqueous Media
By V. Tamara Perchyonok
© V. Tamara Perchyonok 2010
Published by the Royal Society of Chemistry, www.rsc.org

- *Design less hazardous chemical syntheses*: Design syntheses to use and generate substances with little or no toxicity to humans and the environment.
- *Use renewable feedstocks*: Use raw materials and feedstocks that are renewable rather than depleting. Renewable feedstocks are often made from agricultural products or are the wastes of other processes; depleting feedstocks are made from fossil fuels (petroleum, natural gas or coal) or are mined.
- *Use catalysts, not stoichiometric reagents*: Minimise waste by using catalytic reactions. Catalysts are used in small amounts and can carry out a single reaction many times. They are preferable to stoichiometric reagents, which are used in excess and work only once.
- *Avoid chemical derivatives*: Avoid using blocking or protecting groups or any temporary modifications if possible. Derivatives use additional reagents and generate waste.
- *Maximize atom economy*: Design syntheses so that the final product contains the maximum proportion of the starting materials. There should be few, if any, wasted atoms.
- *Use safer solvents and reaction conditions*: Avoid using solvents, separation agents or other auxiliary chemicals. If these chemicals are necessary, use innocuous compounds.
- *Increase energy efficiency*: Run chemical reactions at ambient temperature and pressure whenever possible.
- *Design chemicals and products to degrade after use*: Design chemical products to break down to innocuous substances after use so that they do not accumulate in the environment.
- *Analyse in real time to prevent pollution*: Include in-process real-time monitoring and control during syntheses to minimise or eliminate the formation of by-products.
- *Minimize the potential for accidents*: Design chemicals and their forms (solid, liquid or gas) to minimize the potential for chemical accidents, including explosions, fires and releases to the environment.

Thus, green chemistry involves the study of the removal of these risks fundamentally during the preparation and isolation of chemical materials, based on molecular chemistry; it is not, therefore, the treatment of symptoms.[1] What it does do is replace solvents and reagents with safer ones.

The areas for the development of green chemistry have been identified as follows:

- *Use of alternative feedstocks*: The use of feedstocks that are both renewable rather than depleting and less toxic to human health and the environment.
- *Use of innocuous reagents*: The use of reagents that are inherently less hazardous and are catalytic whenever feasible.

In later chapters, we will see the evolution of free radical chemistry, starting from the typical Bu_3SnH radical hydrogen donor in benzene to the use of broad-range non-toxic and effective hydrogen donors in water and/or aqueous media. Generally, radical reactions with Bu_3SnH initiated by azobisisobutyronitrile (AIBN) proceed effectively in benzene, which bears a conjugated π-system. It is proposed that the radicals formed are stabilised somewhat through the SOMO–LUMO or SOMO–HOMO interaction between the radical and benzene.[3]

Occasionally, it may be required to study the fundamental radical reactions with organotin and benzene. However, the use of radical reactions with such toxic reagents and solvents cannot be considered in the chemical and pharmaceutical industries, even if the results in terms of organic synthesis are excellent and effective. Hence free radical chemists should develop ways to conduct new and less toxic radical reactions in ways that address the 12 principles of green chemistry. Therefore, by creating a new direction in free radical chemistry, green free radical chemistry brings the environmentally benign aspects of free radical chemistry into the spotlight.

Fortunately, radicals are a neutral species in general. Hence they are not generally affected by the various kinds of solvents (reaction media), *i.e.* protic polar solvents such as ethanol and water, aprotic polar solvents such as acetonitrile and dimethyl sulfoxide and non-polar solvents such as hexane and benzene. Moreover, radicals are not affected fundamentally by basic species or acidic species. Radical reactions should take place not only in benzene, but also in water and proceed not only in 1 M aqueous HCl solution, but also in 1 M aqueous NaOH solution. This is the fundamental character of radicals and radical reactions and is a great advantage – an advantage that should be reflected in green chemistry.[4]

Let us look at the history of the development of radical and green chemistry as both of these branches of chemistry were created through a significant paradigm shift of ideas, thoughts and people who trusted in intuition and wanted to be at the frontier of development of knowledge.

1.1 Radicals: Historical and Practical Importance

1.1.1 What Are Free Radicals

Radicals have an impact on all of our lives. We make them in our bodies, they are produced when we light a fire or drive a car and we use plastics as part of our daily living that are produced on a large scale using radical reactions. Radicals affect our health and vitality and govern our ageing through the formation of various harmful radicals. Destructive radicals also affect our environment and radicals generated from chlorofluorocarbons (CFCs) are responsible for the destruction of the Earth's protective ozone layer. So what are free radicals?

We can define radicals as atoms or compounds that contain an unpaired electron. They all contain an odd number of electrons. The single unpaired electron for each atom is represented in formulae by a dot. Almost all radicals can be described as 'free radicals' as they exist independently, free of any support from other species. They are generally very unstable and are regarded as reactive intermediates, together with carbocations, carbanions and carbenes. The high reactivity of radical species is due to the unpaired electron, which would like to pair with a second electron to produce a filled outer shell. The driving force in each case is the formation of a two-electron covalent bond. If we want to prepare the original radicals from the newly formed molecule, we could consider the reverse processes and attempt to break the covalent bond by applying energy (*e.g.* heat or light). This type of bond cleavage, to give each atom one electron, is known as homolysis or homolytic bond cleavage.

There are various pathways by which radicals can react to form stable molecules. They can be combined with themselves or other radicals, but they can also be oxidized to a cation (by loss of an electron) or reduced to an anion (by addition of an electron). These ions could then react with nucleophiles or electrophiles, respectively, to produce neutral and stable products. This is illustrated for a carbon–centred radical in Scheme 1.1. These radicals contain seven valence electrons, which is one electron more than in carbocations and one electron less than in carbanions. We can see that both the radical and cations are electron deficient as they require the addition of one and two electrons, respectively, to produce a filled (eight-electron) outer shell.

Not all radicals are highly reactive. Notably, naturally occurring exceptions include oxygen (O_2) and nitrogen monoxide ($NO^{\bullet}$).[5] Molecular oxygen (O_2) can be thought of as a di- or biradical as it contains two unpaired electrons. Whereas (mono/uni)radicals have an odd number of electrons, biradicals have an even number and O_2 has 12 electrons. The following sections outline the brief history of evolution of free radical chemistry and green chemistry and subsequently discuss a variety of biologically and industrially important radical reactions with a particular focus on the green chemistry aspects of transformations.

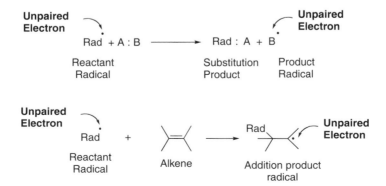

Scheme 1.1   Radicals and reactions in general.

1.1.2 Brief History and Development of Free Radical Chemistry and Evolution of Green Chemistry

In chemistry, as in any other branch of science, experiments often precede theory. Accidental discoveries bring fresh light and reveal aspects hitherto unsuspected of the subject. Neither radical chemistry nor green chemistry is an exception. It is therefore useful, perhaps, to outline, in a very broad manner, the way in which the field developed over the past century or so, viewed from the perspective of the synthetic organic chemist.

In the early days of organic chemistry, when structural and mechanistic concepts were still elusive, it was noted that reactions that could in principle lead to a trivalent carbon species produced dimers instead.

- *1789*: Lavoisier first used the term 'radical' when he described acids as being composed of oxygen and an entity called a radical. The meaning of the name has changed since then but the word 'radical' has remained.[6]
- *1849*: Kolbe described the product derived from the electrolysis of potassium acetate (ethanoate) as a 'methyl radical' with the formula $\cdot C_2H_3$. We now know that the product is ethane (C_2H_6), which is actually a product of dimerization of two methyl radicals ($\cdot CH_3$).[6]
- *1847*: Faraday first demonstrated that oxygen is drawn into a magnetic field and hence is strongly paramagnetic, whereas nitrogen monoxide is weakly paramagnetic. We now associate paramagnetism with molecules (or ions) that contain unpaired electrons, as the spinning electrons behave like tiny magnets and the molecules are drawn into a magnetic field.[7]
- *1900*: Victor Meyer showed that an iodine molecule (I_2) could dissociate into iodine atoms or radicals ($I\cdot$). However, the key breakthrough came in 1900, when Gomberg investigated the reaction of triphenylmethyl bromide with silver. In the absence of oxygen, the reaction yielded a highly reactive white solid which, when dissolved, gave a yellow solution. Gomberg proposed that the product was hexaphenylethane, which, when in solution, existed in equilibrium with coloured triphenylmethyl radical. The presence of a radical helped to explain why other radicals, including oxygen, react rapidly with the product. This was a major discovery and chemists started to accept radicals as a novel chemical entity separate from cations or anions.[8]
- *1911*: The experimental evidence for the formation of the triphenylmethyl radical was so overwhelming that the case for free radicals was firmly established. So why was Gomberg able to observe this particular radical? Part of the reason for the stability of the triphenylmethyl radical can be attributed to the presence of three bulky benzene rings that effectively shield the central carbon atom bearing the radical and slow any reactions.[9]
- *1929*: Paneth showed that tetramethyllead ($PbMe_4$) produces a metallic lead mirror when heated to high temperature (around 200 °C) in a glass tube containing a stream of unreactive carrier gas.[10]
- *1939*: Radicals are postulated to be important intermediates in a variety of chemical reactions, *e.g.* the Kharasch reaction and mechanism,

anti-Markovnikov addition of HBr to alkenes and the first demonstration of radical chain reaction – initiation, propagation and termination nomenclature were introduced. Radical polymerization was introduced, investigated and adapted for industrial use. All these investigations led to the same conclusion that not all carbon-centred radicals are the same: they have a different character and as a result can react differently. This is an important concept, as it allows us to explain and also predict selective radical reactions.[11]

- *1950–70*: Physical organic chemists began to measure and characterise free radicals in order to quantify radical reactions and determine absolute rates of reaction in solution. This was revolutionised by the development of a new technique, known as electron spin resonance (ESR) spectroscopy, which offered a sensitive method for the detection and identification of radical species. The important structural and reactivity features of free radicals have been uncovered through this technique. It was around this time that the synthetic utility of Bu_3SnH as a powerful reducing agent came to light and established the important role of free radical chemistry. It was also around that time that researchers began to investigate and propose mechanisms for the deterioration of fats, oils and other foodstuffs in the presence of oxygen (early examples of *in vitro* models of biologically and environmentally important processes). Radical intermediates were shown to be involved and the term autoxidation was introduced to describe these processes. This prompted the design of molecules, called antioxidants or inhibitors, which could slow or even prevent undesired reactions.[12]

Armed with the knowledge of reaction rates, chemists were now at the point of beginning to explore the use of radicals in the preparation not only of polymers but also of small molecules.[13] Since 1970, a number of important radical reactions have been developed and numerous target molecules have been prepared using these methods.[14] Today, when planning the synthesis of even very complex molecules, the arsenal of chemists is rich with a broad range of free radical reaction classes and also solvents suitable for all kinds of transformations with a great degree of stereo- and even enantio-control that can have a number of advantages over traditional ionic methods (involving anions and cations).[15] It should also be highlighted that radical reactions do not occur only in laboratories. The process by which benz-aldehyde is oxidized in air to benzoic acid on a laboratory bench is the same type of radical reaction as that which leads to the deterioration of foods, the ageing of unprocessed natural rubber and the drying of paints and varnishes. The following sections aim to show the important developments and highlight the origins of free radical chemistry while dispelling the myth and ill-deserved reputation that such molecules are unruly and are likely to produce tar.

The period 1960–70 was an exciting time and also very challenging as environmental concerns became an integral part of public perception of the chemical industry and chemical science in general and the effects that chemicals have on the environment. The foundation of green chemistry was established. Below some important historical dates relating to green chemistry being established as an independent but not mutually exclusive branch of chemistry are outlined.

From the late 1960s to 1970, the environment received a great deal of attention, including formation of the US Environmental Protection Agency (EPA) and the celebration of the first Earth Day, both of which occurred in 1970. In the intervening years, in excess of 100 environmental laws have been passed. These include several major laws listed below:

1. 1970 Clean Air Act, to regulate air emissions.
2. 1972 National Environmental Policy Act, which requires in part that the EPA reviews environmental impact statements for proposed major federal projects (such as highways, buildings, airports, parks and military complexes).
3. 1972 Clean Water Act. This established the sewage treatment construction grants programme and a regulation and enforcement programme for discharges of pollutants into US waters. The Federal Insecticide, Fungicide and Rodenticide Act governs the distribution, sale and use of pesticide products. All of the pesticide products must be registered (licensed) by the EPA. The Ocean Dumping Act regulates the intentional disposal of materials into ocean waters.
4. In 1991, green chemistry became a formal focus of EPA. Green chemistry or environmentally benign chemistry is the design of chemical products and processes that reduce and eliminate the use and generation of hazardous substances.

Green chemistry has gained a strong foothold in the areas of research and development in both industry and academia. Green chemistry means environmentally friendly organic synthesis. The essential aims are to reduce the amount of dangerous, toxic starting materials and by-products (waste disposal) and to reduce damage to the natural environment (Figure 1.1).

1.2 Natural Radical Reactions and Applications

Excessive exposure to environmental pollution (*e.g.* exhaust fumes), ultraviolet light or cigarette smoke and illness can cause the body to produce harmful radicals. It has been estimated, for example, that 10^{14} radicals, which include $NO^{\bullet}$ and $NO_2^{\bullet}$, are present in one puff of cigarette smoke. Left unchecked, destructive radicals can lead to a number of diseases in humans, including arthritis, cancer and Parkinsonism. Radicals may also start the damage that causes fatty deposits in the arteries, leading eventually to heart disease or a stroke, and experimental work points to the role of radicals in bovine spongiform encephalopathy(BSE).

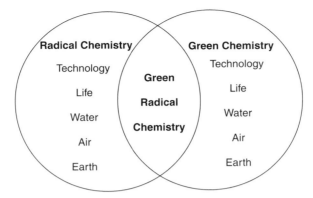

Figure 1.1   Green free radical chemistry evolving from free radical chemistry and green chemistry.

A role for radicals in ageing was first suspected in the 1950s and recent research has attributed this to oxygen-centred radicals derived from mito-chondria. The mitochondrion is the molecular energy factory of cells where oxygen and nutrients are used to prepare adenosine triphosphate (ATP) – the molecule that powers most other activities in cells. Unfortunately, one by-product of these processes is superoxide radical anion ($O_2^{-\bullet}$), a charged radical, which, in turn, can be converted to hydroxyl radical ($HO^\bullet$), the most reactive oxygen radical known. The hydroxyl radical is particularly indiscriminate in its choices of reactant and can damage proteins, fats and deoxyribonucleic acid (DNA) within the cell. These reactions can interfere with the proper func-tioning of the cell, leading to its death and ultimately that of the organism. Radicals of this type are also produced when water is irradiated with high-energy radiation (e.g. ultraviolet light, X-rays or γ-rays). It also important to note that water accounts of 70% of our bodyweight. Exposure to sunlight, X-ray irradiation or nuclear radiation can lead to the formation of these destructive radicals in our bodies, which, in turn can promote cancers, sterility and even death. Hydroxyl radical generation is actually used beneficially to treat cancer with radiotherapy: the rapidly multiplying cancer cells are killed by exposure to a radioactive materials (e.g. ^{60}Co) or X-rays.

It is wrong, however, to believe that our body is completely defenceless against damaging radicals (Figure 1.2). Living cells can produce enzymes which detoxify free radicals, including superoxide dismutases (SODs). These are the metal-containing enzymes (e.g. manganese, copper and zinc) that convert O_2^{-} to oxygen and hydrogen peroxide. The reactive peroxide is readily reduced to water by catalase or peroxidase enzymes. β-Carotene (pro-vitamin A), vitamin C (ascorbic acid) and vitamin E (α-tocopherol) can also delay or inhibit oxidative damage and these types of molecules are known as antioxidants (Figure 1.3). Vitamin E is fat soluble and protects against radical damage within the cell membrane; laboratory experiments suggest that this process also involves the water-soluble vitamin C.

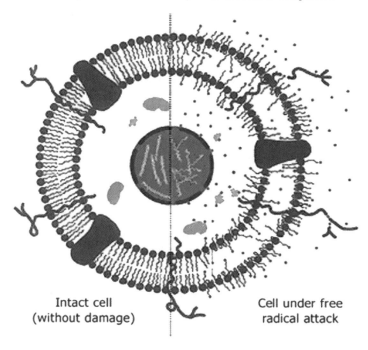

Figure 1.2 Intact cell and cell under free radical attack.

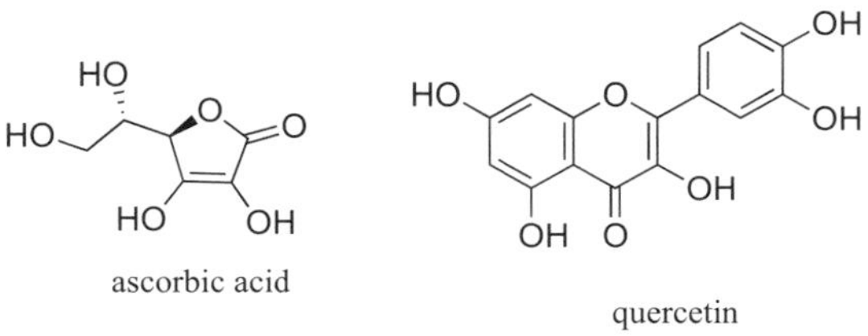

Figure 1.3 Common antioxidants.

These molecules scavenge radicals and vitamin E can react with an alkoxyl radical by donating a hydrogen atom (from the OH group) to give an alcohol and a vitamin E radical, which is much less harmful. Vitamin E is believed to be regenerated by reaction with vitamin C at the cell membrane (water–lipid) surface.

1.3 Autoxidation in Biology: a Brief Chemist's Perspective

Our antioxidation defences therefore rely on both vitamins and minerals, from which metal-containing enzymes are prepared, in our diet. A variety of alternative antioxidants are present in food and include flavonoids, such as quercetin. Flavonoids are a group of around 1000 compounds found in certain fruits and vegetables, including citrus fruit, apples and grapes. Tea leaves are also full of flavonoids, and garlic produces a powerful antioxidant called allicin. Just how important these dietary antioxidants are to our health is a matter of debate. Recent research suggests, however, that antioxidants may offer protection against cancer and heart disease. Hence could it be that the antioxidants in apples can really keep the doctor away?

Not all radical reactions involving molecular oxygen are destructive in humans. A very important series of beneficial reactions arises from the reaction of oxygen with an unsaturated acid called arachidonic acid. These radical reactions, often termed the 'arachidonic acid cascade', produce a variety of biologically active compounds (including prostaglandins, thromboxanes and leukotrienes) that are vital for the body to function properly. Of particular medicinal importance are prostaglandins, including prostaglandin F_{2a}, which exhibit a range of important activities, including contraction of smooth muscles, regulation of cell function and fertility, lowering of arterial blood pressure and control of platelet aggregation (Scheme 1.2).

One final example of natural product radicals involves the enediyne family of compounds isolated from various bacteria.[16] These are some of the most potent antitumour agents ever discovered and dynemicin A, for example, is known to prolong significantly the lifespan of mice inoculated with leukaemia cells. The remarkable biological activity of these compounds results from their reaction with DNA in the rapidly multiplying cancer cells. Enediyne antitumour agents are able to generate a very reactive biradical intermediates, known as 'warheads', that can selectively abstract two hydrogen atoms from the DNA backbone (Scheme 1.3). This leads to irreversible cleavage of the DNA strand, which ultimately leads to cell death.[17]

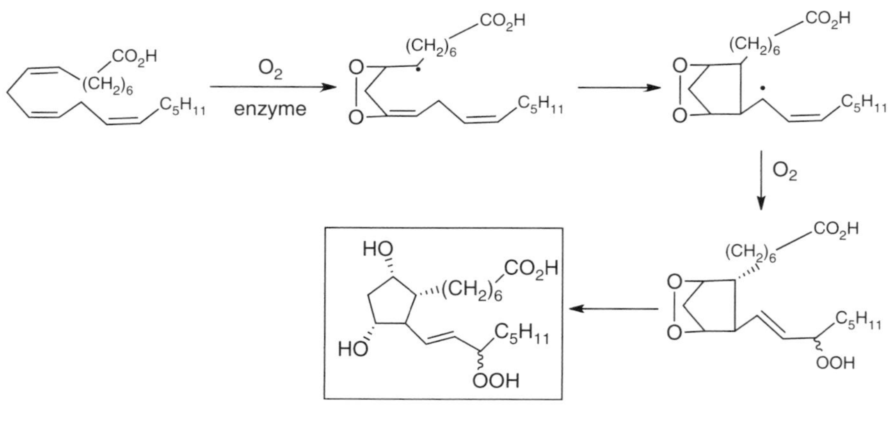

Scheme 1.2   Arachidonic acid cascade and prostaglandin mechanism.

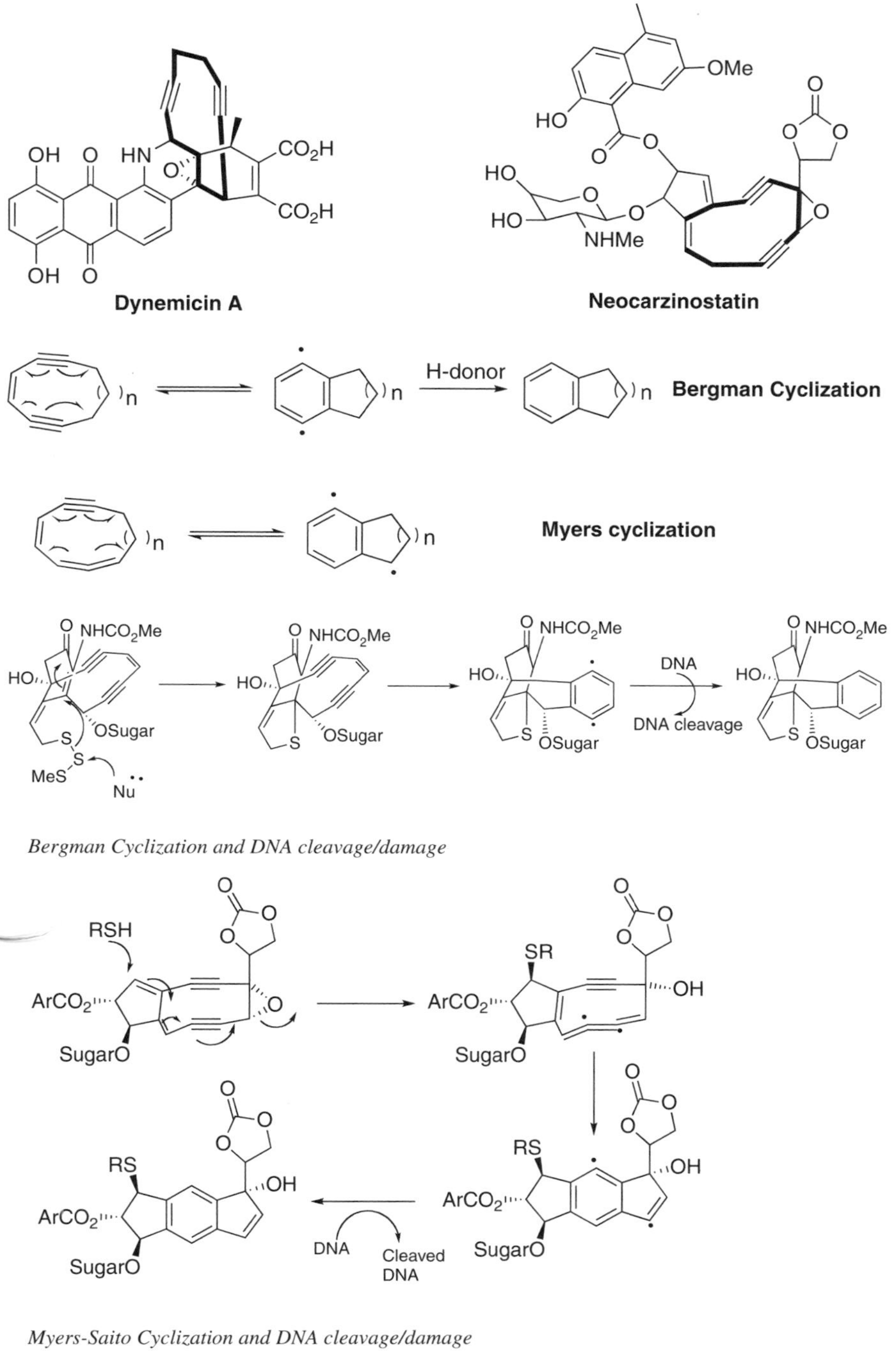

Dynemicin A

Neocarzinostatin

Bergman Cyclization

Myers cyclization

Bergman Cyclization and DNA cleavage/damage

Myers-Saito Cyclization and DNA cleavage/damage

Scheme 1.3 Bergman and Myers–Saito cyclizations and DNA cleavage/damage.

1.4   Commercial Radical Reactions

The most important industrial applications of radical reaction to date are used for the manufacture of polymers. Around 10^8 tonnes (or 75%) of all polymers are prepared using radical processes. These are chain reactions in which an initial radical adds to the double bond of an alkene monomer and the resulting radical adds to another alkene monomer and so on. This addition polymerisation is used to make a number of important polymers, including poly(vinyl chloride) (PVC), polystyrene, polyethylene and poly(methyl methacrylate). Copolymers can also be easily prepared starting from a mixture of two or more monomers. These polymers have found widespread use as they possess a range of chemical and mechanical properties (such as strength and toughness).

A related polymerisation reaction is responsible for drying paints, varnishes and oils. Oil-based varnishes contain mixtures of long-chain fatty acids such as linoleic acid. When exposed to air, the acids react with oxygen to form various oxygen- and carbon-centred radicals. These can then react with the double bonds of nearby molecules to 'cross-link' (or join) the acids and this eventually leads to the formation of a solid polymer. Although these paints are cheap, they are slowly falling into disuse, owing to their poor drying properties and also continuous reaction of the aerial oxygen that leads to a yellow and brittle appearance of the surface, a process known as weathering. More modern emulsion (or latex) paints contain a suspension of the polymer in water which quickly evaporates to leave a coating.

Combustion is another very important process which we encounter every day. The car engine is a 'radical chemist at heart': it exploits the reaction of petrol with molecular oxygen at high temperature. Although the mechanism of combustion of hydrocarbons is complex, many of the steps involve the production and reaction of carbon- and oxygen-centred radicals. If the production of the radicals is not checked, an explosion or 'knock' can occur. To prevent this problem, an anti-knock agent, such as $PbEt_4$ or, more recently, methyl *tert*-butyl ether ($MeOCMe_3$), is added to petrol. $PbEt_4$ is a source of ethyl radicals and these can combine with the hydrocarbon radicals, stopping any further radical reactions and preventing the explosion.

Similar hydrocarbon radicals are generated in industrial plants devoted to 'cracking'. The high temperature ($\geq 600\,^{\circ}C$) ensure that radicals are produced from homolytic cleavage of the strong carbon–carbon and carbon–hydrogen bonds. This process allows valuable alkenes (such as ethene) and short-chain hydrocarbons to be prepared by degrading long-chain hydrocarbons extracted from crude oil.

Molecular oxygen can also oxidize a variety of organic compounds, including hydrocarbons, aldehydes, amines, ethers and ketones. These auto-oxidation reactions can be used to make a variety of small molecules and a number of industrial processes rely on the controlled oxidation of organics using molecular oxygen (often with a metal catalyst). Examples include the formation of phenol and acetone from cumene (isopropylbenzene) and cyclo-hexanone from cyclohexane. Phenol is a popular starting material for a number of other products, including aspirin and phenolic resins, and cyclohexanone can be oxidised (using nitric acid) to adipic acid, a precursor of nylon.

However, not all radical reactions with molecular oxygen and organics are desirable. The widely known and notable example is reaction with unsaturated fatty acids; oxygen can react with double bonds to produce ultimately a complex mixture of volatile, rancid-smelling, short-chain carboxylic acid products. This autoxidation process can lead to rancid butter and particular margarines which are rich in polyunsaturates. However, the radical chain reaction involved in the oxidation processes can be prevented by sealing foodstuffs under nitrogen (inert atmosphere) or adding antioxidants, such as vitamin E, to foods. Other commonly used additives include butylated hydroxytoluene and butylated hydroxyanisole.

The chapter highlights how far radical chemistry has come from the initial perception of being too reactive and useful for nothing more than the production of tar to a foundation science behind many important biological and industrial processes. The intersection of green chemistry and radical chemistry leading to green free radical chemistry will be expanded upon in more detail in later chapters.

References

1. P. T. Anastas and J. C. Warner, *Green Chemistry and Practice*, Oxford University Press, Oxford, 1998.
2. P. A. Hamley and M. Poliakoff, *Chem. Eng.*, 2001, **72**, 24; S. K. Ritter, *Chem. Eng. News*, 2001, **79** (29), 24.
3. B. Giese, *Radicals in Organic Synthesis: Formation of Carbon–Carbon Bonds*, Pergamon Press, Oxford, 1986.
4. C. Walling, *Fifty years of free radical chemistry. Chem. Br.*, 1987, **23**, 767–770.
5. Z. Alfassi, *The Chemistry of Free Radicals: N-Centered Radicals*, Wiley, New York, 1998.
6. M. Berthelot, *La Révolution Chimique: Lavoisier*, Alcan, Paris, 1890; J. B. Dumas, *M. Lavoisier, Théoricien et Expérimentateur*, Presses Universitaires de France, Paris, 1955; A. Lavoisier, *Traité lémentaire de Chimie, Présenté dans un Ordre Nouveau et d'Après les Découvertes Modernes*, 2 vols, Chez Cuchet, Paris, 1789, Reprinted Cultures et Civilisations, Brussels, 1965.
7. H. M. Leicester, *The Historical Background of Chemistry*, Dover Publications, New York, 1971. (Reprint of 1956 text first published by John Wiley and Sons, Ltd.).
8. M. Gomberg, Radicals in chemistry, past and present, *Ind. Eng. Chem.*, 1928, **20**(2), 159.
9. M. Gomberg, *J. Am. Chem. Soc.*, 1900, **22**, 757; M. Gomberg, *Chem. Ber.*, 1900, **33**, 3150.
10. F. Paneth and W. Hofeditz, *Chem. Ber.*, 1929, **62**, 1335.
11. M. S. Kharasch, E. T. Margolis and F. R. Mayo, *J. Org. Chem.*, 1937, **2**, 393; F. R. Mayo and F. M. Lewis, *J. Am. Chem. Soc.*, 1944, 66, 1594; F. R. Mayo, F. M. Lewis and C. Walling, Discuss. *Faraday Soc.*, 1947, 2, 285.

12. W.A. Waters, *The Chemistry of Free Radicals*, Clarendon Press, Oxford, 1946; C. Walling, *Free Radicals in Solution*, Wiley, New York, 1957.

13. J. K. Kochi (Ed.), *Free Radicals*, Wiley, New York, 1973.

14. H. Fischer (Ed.), *Landoldt–Bornstein, New Series*, **Vol. 13**, Springer, Berlin, 1983.

15. D. H. R. Barton and W. B. Motherwell, in *Organic Synthesis Today and Tomorrow*, ed. B. M. Trost and C. R. Hutchinson, Pergamon Press, Oxford, 1981, p. 1; D. J. Hart, *Science*, 1984, 223, 883; B. Giese, *Angew. Chem. Int. Ed. Engl.*, 1985, **24**, 553; B. Giese (ed.), *Selectivity and Synthetic Applications of Radical Reactions*, Tetrahedron 'Symposia in Print', *Tetrahedron*, 1985, **41**, 3887; N. A. Porter and M. O. Funk, *J. Org. Chem.*, 1975, **40**, 3614.

16. K. C. Nicolau and W. -M. Dai, Chemistry and biology of the enediyne antibiotics, *Angew. Chem. Int. Ed. Engl.*, 1991, **30**, 1387.

17. R. R. Jones and R. G. Bergman, p-Benzyne. Generation as an intermediate in a thermal isomerization reaction and trapping evidence for the 1,4-benzenediyl structure, *J. Am. Chem. Soc.*, 1972, **94**, 660–661.

CHAPTER 2

Basic Radical Chemistry: General Aspects of Synthesis with Radicals

Radicals are species with at least one unpaired electron, which, in contrast to organic anions and cations, react easily with themselves in bond-forming reactions. In the liquid phase, most of these reactions occur with diffusion-controlled rates. Radical–radical reactions can be slowed only if radicals are stabilized by electronic effects (stable radicals) or shielded by steric effects (persistent radicals). However, these effects are not strong enough to prevent diffusion-controlled recombination of, for example, benzyl radicals or *tert*-butyl radicals.[1] Only in extreme cases are the radical or di-*tert*-butylmethyl radical recombination rates low.[2] While the recombination rates of the triphenylmethyl radical is reduced due to both steric and radical stabilizing effects, the steric effect alone slows the recombination of the di-*tert*-butylmethyl radical. Since neither of the radicals have C–H bonds β to the radical centre, disproportionation reactions, in which the hydrogen atom is transferred, cannot occur.

2.1 Reactions Between Radicals

The fact that reactions between radicals are in most cases very fast could lead to the conclusion that direct radical combination is the most synthetically useful reaction mode.[3] This, however, is not the case because direct radical–radical reactions have several disadvantages:

- In the recombination reactions, the radical character is destroyed so that one has to work with at least an equivalent amount of radical initiators.
- The diffusion-controlled rates in radical–radical reactions give rise to low selectivity which cannot be influenced by the reaction conditions.

RSC Green Chemistry No. 6
Radical Reactions in Aqueous Media
By V. Tamara Perchyonok
© V. Tamara Perchyonok 2010
Published by the Royal Society of Chemistry, www.rsc.org

- The concentrations of radicals are so low that reactions with non-radicals, such as solvents, which are present in high concentrations are very difficult to prevent.

2.2   Reaction Between Radical and Non-radical

Nevertheless, there are several classes of synthetically useful reactions involving free radicals with non-radicals.[3] They possess the following advantages:

- The radical character is not destroyed during the reaction; therefore, one can work with catalytic amounts of radical initiators.
- Most of the reactions are not diffusion controlled and the selectivities can be influenced by variation of the substituents.
- The concentration of the non-radicals can be easily controlled.

In most cases, in order to apply reactions between radicals and non-radicals in synthesis, chain reactions have to be encouraged and established. For the successful use of radical chains, two fundamental conditions have to be met:

- The selectivities of the radicals involved in the chain have to differ from each other.
- The reaction between radicals and non-radicals must be faster than radical combination reactions.

In practice, these rules can best be illustrated by a chain reaction that has gained increasing synthetic application over the years and become one of the fundamental pillars of free radical chemistry.[3] In this chain reaction, alkyl halides and alkenes react in the presence of tributyltin hydride to give products.

For the successful application of this tin method, alkyl radicals must attack alkenes to form adduct radicals. Trapping of the newly formed radical leads to the formation of the products and tributyltin radicals, which react with alkyl halides to give back educt radicals. The tin method can be synthetically useful only if these reactions are faster than all other possible reactions of the radicals formed. Therefore, the radicals in the chain must meet certain selectivity and reactivity prerequisites.

2.3   Reactivity and Selectivity

As was previously mentioned, radicals can undergo a number of different and competitive reactions. These processes have different rates of reaction and if one reaction proceeds at a much faster rate than all the others we have a selective and high-yielding process. Alternatively, if a variety of reactions proceed at similar rate, the radical will react unselectively to produce a number of different products. The rates of these reactions can vary enormously and, for example, the rate constants of abstraction reactions can vary by a factor of at

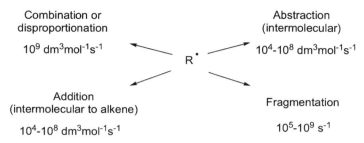

Scheme 2.1   Radicals and scope in synthesis.

least 10 000. The key factors that influence radical reactivity include enthalpy, entropy, steric effects, stereoelectronic effects, polarity and redox potential.[4] Scheme 2.1 represents scope and kinetic parameters for synthetically useful free radical transformations.

2.4   Enthalpy: in Brief

Radical reactions will generally proceed so as to convert a radical into a more stable radical or non-radical product. Whereas combination reactions leading to non-radical products have similar (diffusion-controlled) reaction rates, the rates of radical–non-radical reactions can vary enormously. As a guide, the more reactive the radical reactant and the more stable the product, the faster is the reaction rate. In practical terms, this explains why the reactive phenyl radical abstracts a chlorine atom much faster from carbon tetrachloride than does the *tert*-butyl radical (in solution at 25 °C). Reaction with the phenyl radical is favoured because the carbon–chlorine bond in chlorobenzene is stronger than that in 2-chloro-2-methylpropane. We can therefore predict whether a radical reaction will take place by considering the energies of the bonds that are broken and those that are formed. This will provide an approximate enthalpy change (ΔH°) for the reaction; if energy is released, the reaction is exothermic ($-\Delta H^{\circ}$); if energy is absorbed, the reaction is endothermic (ΔH°). Exothermic reactions result in the formation of strong bonds and these can proceed rapidly (often spontaneously), whereas endothermic reactions (which lead to products with weaker bonds) are generally very slow[3,5] (Scheme 2.2).

2.5   Entropy

The enthalpy of a reaction gives only an approximate guide to selectivity in radical reactions, because the Gibbs free energy equation shows that temperature and entropy are also important factors in determining the outcomes of a reaction. For a reaction to be thermodynamically favoured, ΔG° should be negative; this occurs when there is a negative enthalpy change (ΔH°), a positive

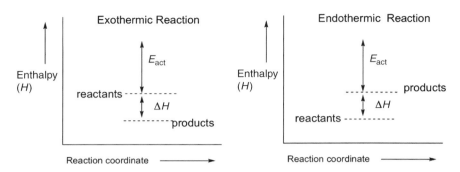

Scheme 2.2   Reaction profile for exothermic and endothermic transformations.

entropy change (ΔS°) and a high temperature (T). Therefore, reactions that produce an increase in entropy (or disorder) by increasing the number of molecular species on going from reactants to products are favoured. This explains why some endothermic reactions (positive ΔH°) with an increase in entropy (positive ΔS°) do not proceed spontaneously at room temperature. Higher temperatures are required so as to increase the $T\Delta S^{\circ}$ contribution (above that of ΔH°) to give a negative value of ΔG°.

$$\Delta G = \Delta H - T\Delta S \qquad \text{with } -\Delta G \text{ when } T\Delta S > \Delta H$$

It is probably appropriate to link a few common concepts of a radical reaction such as radical initiators, reaction types and the driving force for reactions in the context of entropy to make the link between the physical organic and free radical chemistry even stronger.

The decomposition of peroxides (RO–OR) and azo compounds (R–N=N–R) to form radicals is favoured by an increase in entropy. One peroxide molecule decomposes to give two alkoxyl (RO$^{\bullet}$) radicals, whereas an azoalkane will form three species: two carbon-centred radicals (R$^{\bullet}$) and nitrogen gas. The formation of the gas is particularly favourable because of the greater freedom of motion in gases (than in liquids or solids), resulting in an increase in disorder.

The driving force for a number of fragmentation reactions, which are often endothermic, is an increase in entropy. An example of such transformations is the decarbonylation reaction of acyl radicals [RC(=O)$^{\bullet}$], which is generally endothermic, but can proceed with reasonable rates (10^{4}–10^{7} s^{-1} at 25 °C). In addition to these reactions being exothermic, they are also favoured by entropy and both of these factors contribute to give a fast and irreversible decarboxylation reaction.

For radical rearrangement reactions, a single reactant leads to a single product with nearly the same mass and an almost identical structure. Although cyclisation reactions can be categorised as rearrangements, they generally show

a decrease in (rotational) entropy, as bond rotation or the number of conformations in the cyclic product is not as great as that for the open-chain
starting materials. The origin of favoured *versus* unfavoured radical cyclizations can be explained using these fundamental principles and is applicable to a
broad range of cyclisations involving small, medium and large ring formations.
Another important class of free radical transformations is the intermolecular
hydrogen atom abstractions (S_Hi), which are also controlled by entropy and the
linear geometry required in the transition state can be achieved in 1,5- and
higher abstraction reactions. However only 1,5- and 1,6-atom abstractions are
commonly observed because, with longer chains, there is a greater loss of
entropy in the transition state.

A much greater loss of entropy occurs in intermolecular reactions when two
reactants collide to form one product. Although the system has now become
more ordered, for radical–radical reactions this does not have a major impact
on the rate of reaction because combination and disproportionation reactions
are very exothermic. The reactions of radicals and non-radicals appear to be
greatly influenced by entropic effects. These reactions are less exothermic (or
entropy favoured) because a radical rather than a non-radical product is
formed. Entropy now becomes much more important and the rates of these
intermolecular reactions are considerably slower. The entropy factor explains
why intermolecular radical additions can be up to 10^5 times slower than related
intramolecular cyclisations.

2.6 Steric Effects

A negative value for ΔG° tells us that a reaction can take place, but the rate of
the reaction can be determined by the enthalpy of activation (ΔH°). This is a
measure not only of the difference in bond energy between the starting materials and the transition states, but also the difference in bond strain. The more
strained the transition state, the higher is the value of ΔH° and the slower the
reaction. Therefore, even though two reactions can be thermodynamically
favoured and have similar (negative) values for ΔH°, they will not have the
same rate if ΔH° is different. It is a reasonable observation that radical reactions usually have low values for ΔH° because the majority of radicals are
reactive. However, this statement does not hold true for persistent or long-lived
radicals that have very bulky substituents surrounding the radical centre.
Reaction of these sterically hindered radicals would require a very strained
transition state with a high enthalpy of activation and this is disfavoured.

Steric effects are used to explain the regioselective addition of radicals to
alkenes. The radical preferentially attacks the less substituted (hindered) end of
the double bond to give a less strained transition state with lower ΔH° in an
anti-Markovnikov-type reaction. If we introduce substituents on to the double
bond, the rate of addition is lowered because of greater steric interaction. Thus,
for example, the methyl radical will add three times more quickly to ethene
(CH_2CH_2) than to the disubstituted alkene (*E*)-2-butene.

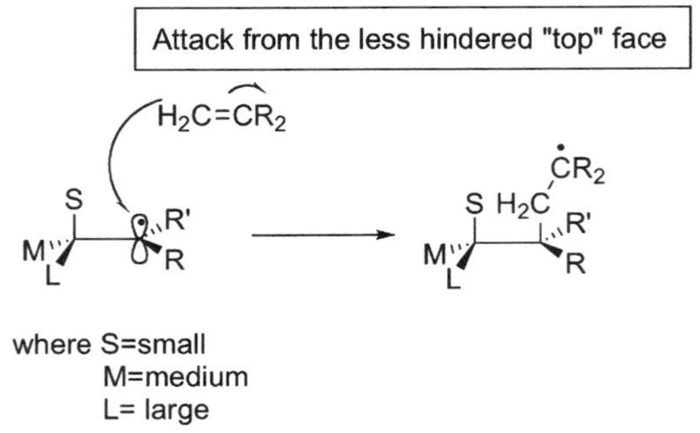

Scheme 2.3   Radical addition reaction and stereospecificity in general.

Steric effects can also explain why carbocyclic radicals (which are not able to undergo free rotation about the C–C bonds) can add to alkenes stereo-selectively. The introduction of an adjacent chiral centre can make the two 'faces' of the radical non-equivalent and this can lead to the alkene pre-ferentially adding to the less hindered face (Scheme 2.3).

2.7   Stereoelectronic Effect

As mentioned previously, radicals with electron-withdrawing or electron-donating substituents can be stabilized by interaction of the singly occupied orbital with an n-, π- or σ-orbital. For effective stabilization, the interacting orbitals must overlay efficiently and this will depend on their geometry or position in space. Similarly, for a radical to react, the singly occupied orbital must be able to overlap with either another of its own orbitals (for intramo-lecular reactions) or another radical or non-radical orbital in a different molecule. For reaction of a radical with the different molecule, there is often no restriction on the orbital geometries and they can rotate freely so as to combine with the maximum overlap.

For an S_H2 reaction, a radical and non-radical can therefore orientate themselves to give a linear transition state which maximizes the interaction of the radical orbitals and the vacant σ-orbital of the bond which is broken. These may, however, be difficult for very bulky molecules, where steric hindrance can prevent the orbitals from becoming close enough for efficient overlap (Scheme 2.4).

For intramolecular radical reactions, there is often a restriction on the orbital geometries. The structure of the starting material dictates how far the orbitals are apart and their relative position; if they are situated a long way apart or held rigidly on different sides of a ring, they will find it difficult to interact. When the radical bonds are not in close proximity to the C–H σ-orbital, intermolecular hydrogen atom abstraction can compete with the intramolecular processes. Even when the orbitals are very close together, they may not interact because their geometry can prevent orbital overlap. This

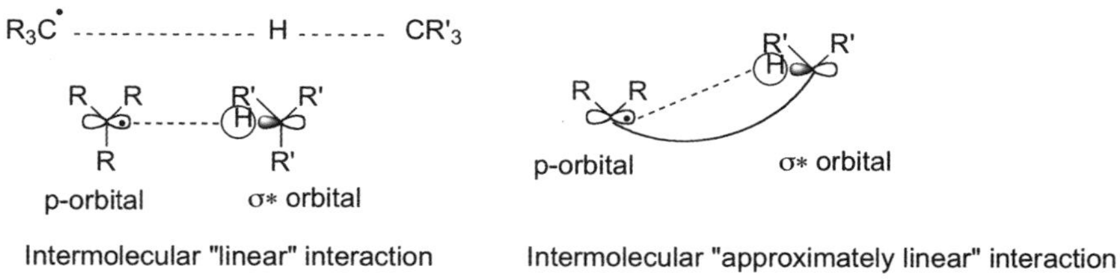

R₃C• -------------- H ------- CR'₃

p-orbital   σ∗ orbital

Intermolecular "linear" interaction

p-orbital   σ∗ orbital

Intermolecular "approximately linear" interaction

Scheme 2.4   Transition state requirements and arrangements for hydrogen atom transfer reactions.

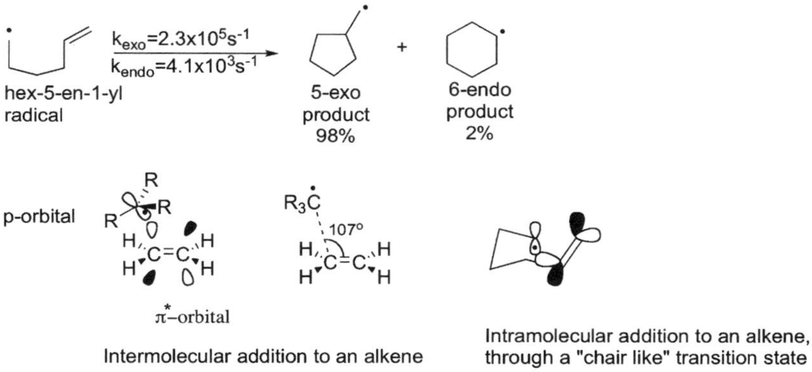

$$k_{exo}=2.3\times10^5 s^{-1}$$
$$k_{endo}=4.1\times10^3 s^{-1}$$

hex-5-en-1-yl radical

5-exo product 98%

6-endo product 2%

p-orbital

π^*-orbital

Intermolecular addition to an alkene

Intramolecular addition to an alkene, through a "chair like" transition state

Scheme 2.5   Radical cyclization reactions: *5-exo vs. 6-endo* and transition state requirements.

so-called electronic strain explains why 1,3- and 1,4-hydrogen atom transfers are very slow. The short chain length restricts the orbital positions and so the radical orbital cannot attack the vacant σ-orbital at an angle of approximately 180° to give the most efficient overlap.

Stereoelectronic factors have been used to explain why the 5-*exo* mode of cyclisation is faster than the competitive 6-*endo* cyclisation. For the hex-5-en-1-yl radical, the rate of 5-*exo* cyclisation is around 60 times faster than that for the 6-*endo* reaction at 25 °C. This is surprising as the 6-*endo* product is a secondary radical and (thermodynamically) more stable than the primary radical formed on 5-*exo* cyclisation. In addition, the five-membered cyclopentane ring is more strained than the cyclohexane ring. The cyclisation is therefore under kinetic control and this is because the singly occupied orbital of the radical attacks the alkene at an angle close to 107° and so can overlap more favourably with the alkene π*-orbital at the 'internal' carbon atom. The carbon chain will position the radical's p-orbital closer to the internal (rather then the external) alkene carbon and the energy calculations have confirmed that the smaller ring is formed because the chair-like transition states are less strained and lower in energy (Scheme 2.5).

Radical fragmentation reactions are also stereoelectronically controlled. For the β-elimination or fragmentation of carbon–carbon centred radicals, the radical p-orbital and the σ-orbital of the C–X bond that is broken must lie in the same plane so that the double bond can be easily formed. This is not a problem for molecules that can rotate freely around the central carbon–carbon bond, as rotation can position the radical and the C–X bond in the correct orientation. The cyclopropylmethyl radical, for example, can undergo rotation around the exocyclic carbon–carbon bond to align the p- and σ-orbitals and very rapid β-elimination occurs to open the strained three-membered ring. In contrast, the cyclopropyl radical cannot undergo ring opening because the carbon–carbon bonds within the ring cannot rotate and so the p- and σ-orbitals lie at 90° (or orthogonal) to one another. This is in spite of the fact that the three-membered ring is very strained and ring opening is thermodynamically favoured.

2.8   Polarity

For reactions to occur, the interacting frontier orbitals must not only interact efficiently, but also have similar energies. The occupied frontier orbital for a radical is called the singly occupied molecular orbital (SOMO), and we have previously discussed that radicals bearing different substituents have different SOMO energies. Radicals adjacent to electron-donating groups interact with a filled orbital to give a high-energy SOMO, whereas radicals next to electron-withdrawing groups interact with an unfilled orbital to give a low-energy SOMO. The SOMO energies lie somewhere between the highest occupied molecular orbital (HOMO) and lowest unoccupied molecular orbital (LUMO) of a non-radical. Therefore, for the reaction of a radical with a non-radical, we need to consider the SOMO–HOMO and SOMO–LUMO interactions. In both cases, the interaction will lead to a decrease in energy and the formation of a bond; a SOMO–HOMO interaction places two of the three electrons in a low-energy bonding orbital, whereas a SOMO–LUMO interaction takes places *via* the single electron in a low-energy bonding orbital. The energy of the SOMO will determine whether interactions with the HOMO or (higher energy) LUMO predominates. Electrophilic radicals (with a low-energy SOMO) will be closer in energy to the HOMO and therefore the SOMO–HOMO interactions will predominate. In comparison, nucleophilic radicals (with a high-energy SOMO) will be closer in energy to the LUMO and therefore the SOMO–LUMO interaction will predominate (Scheme 2.6).

2.9   Elementary Reaction Steps Between Radicals and Non-radicals

A radical chain is built up of different types of propagation steps, all of which lead to the new radicals:

- addition reactions
- substitution (abstraction) reactions

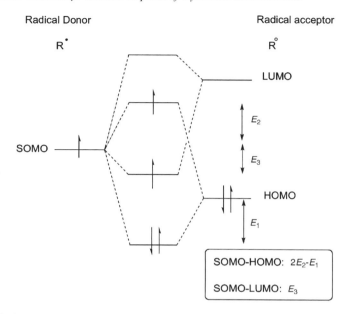

Scheme 2.6   Orbital interactions of unpaired electrons with empty and filled orbitals.

- elimination (fragmentation) reactions
- rearrangement reactions
- electron transfer reactions.

As a rule of thumb, there are two simple rules that can be applied to fast radical chain reactions:

- Most chain-propagating steps are exothermic and one can use the strength of bonds that are broken and formed as a rough guide to the rate of the reaction (thermodynamic parameter).
- Because of the early transition states in fast radical reactions, frontier molecular orbitals theory can be utilized for these reactions.

2.10   Additions

Addition of alkyl radicals to alkenes is a useful C–C bond formation reaction in which a σ-C–C bond is made from a π-C–C bond in a very exothermic reaction. In contrast, π-C–O bonds of ketone and aldehyde are nearly as strong as σ-C–C bonds. Therefore, ketones and aldehydes cannot be used as intermolecular traps in synthesis (Scheme 2.7).

The rate of addition of a radical to an alkene depends largely on the substituents on the radical and the alkenes. This substituent effect can be supported by the frontier orbital series. The singly occupied (SOMO) of the radical

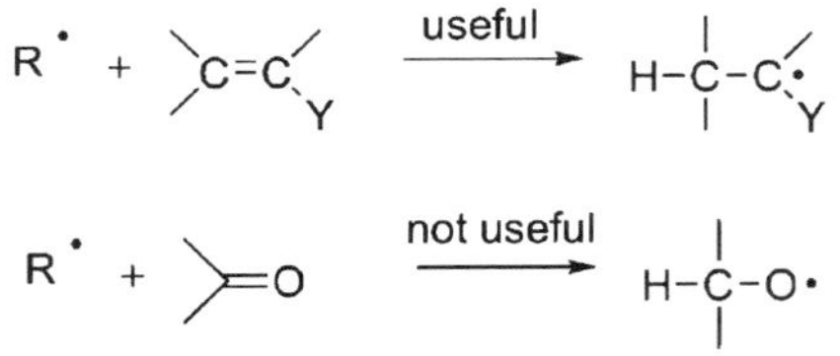

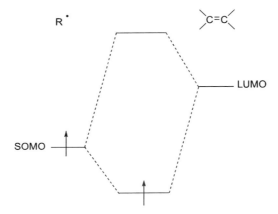

Scheme 2.7   Competition between radical addition to C=C and C=O bonds.

Scheme 2.8   Orbital overlap of R° with the C–C multiple bond.

overlaps/interacts with the lowest unoccupied orbital (LUMO) and/or the highest occupied orbital (HOMO) of the C–C multiple bond. Radicals with a high-lying SOMO interact preferentially with the LUMO of the alkene (Scheme 2.8).

2.11   Orbital Interaction Between a Nucleophilic Radical and an Electron-poor Alkene

Electron withdrawing substituents at the alkene, which lower the LUMO energy, increase the addition rate by reducing the SOMO–LUMO difference. Therefore, cyclohexyl radicals react almost 8500 times faster with acrolein than with 1-hexene.[4]

The orbitals are crucial to the fact that a *tert*-butyl radical reacts faster than primary or secondary radicals with alkenes such as vinylphosphinic esters or acrylonitrile. Thus, the increase in the SOMO energy on going from primary to tertiary radicals has a larger effect on the rate than the decrease in the strength of the bonds that are formed. Alkyl, alkoxyalkyl, aminoalkyl and other similar radicals are therefore nucleophiles. However, radicals with electron-withdrawing substituents at the radical centre have SOMO energies that are so low that the SOMO–HOMO interaction dominates (Scheme 2.9).

2.12 Orbital Interaction Between an Electrophilic Radical and an Electron-rich Alkene

These radicals react like electrophiles, that is, electron-donating substituents on the alkene increase the rate. The malonyl radical, for example, reacts with enamines 23 times faster than with acryl esters.

2.13 Substitution (Abstraction) Reactions

Since an O–H bond is much stronger than a C–H bond, a typical bimolecular reaction for alkoxy radical is H-abstraction from C–H bonds. However, the O–H bond of an alcohol is attacked too slowly for synthetic applications because the reaction is thermoneutral (Scheme 2.10).

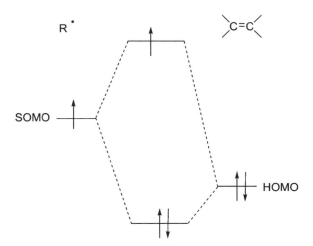

Scheme 2.9   Orbital overlap of R˙ (either nucleophilic or electrophilic) with C–C multiple bonds.

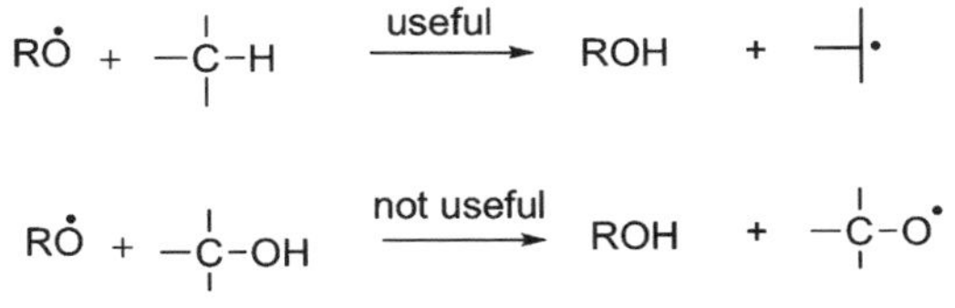

RȮ + —C̦—H →(useful)→ ROH + —|˙

RȮ + —C̦—OH →(not useful)→ ROH + —C̦—O˙

Scheme 2.10   RO˙ radical addition to C–H and C–OH bonds in competition.

Alkoxy radicals are electrophiles and they preferentially attack C–H bonds with high HOMO energies, for instance, the α-C–H bond of ethers and amines or the alkyl C–H bond of esters. In contrast, nucleophilic alkyl radicals abstract a hydrogen atom from the acyl group of esters, because this C–H bond has a lower LUMO energy.

These differences also account for the preferential abstraction of the β-hydrogen of propionic acid by the electrophilic chlorine atom and the abstraction of the α-hydrogen by the nucleophilic methyl radical.

2.14   Elimination

In elimination reactions, two molecules are formed from one. Hence these reactions are favoured by activation entropies and the free energy gain increases with increasing reaction temperature. Therefore, even alkoxy radicals undergo fast β-elimination reactions, although the enthalpy differences between π-C–O and σ-C–C bonds are small. However, a C–OR bond at a radical centre is cleaved too slowly to be of synthetic use, because less stable π-C–C bonds are formed. Only with the weaker C–Br, C–SR or C–SnR$_3$ bonds are β-elimination reactions fast enough to be synthetically useful (Scheme 2.11).

2.15   Rearrangement Reactions

Compared with their cationic counterparts, only a few radical rearrangements are fast enough to appear in synthesis. An example is vinyl migration in which a vinyl group migrates *via* the formation of the methylcyclopropyl intermediate (Scheme 2.12).

This reaction is a combination of an intramolecular addition and an elimination reaction. The addition of the nucleophilic radical to the electron-rich alkene is fast because the decrease in entropy is much smaller than that in

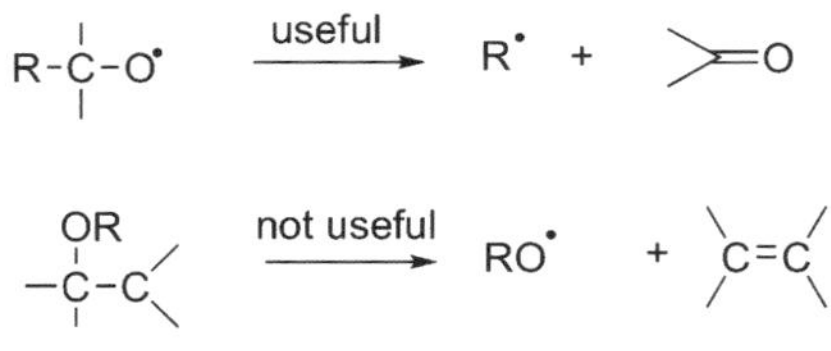

Scheme 2.11   Radical fragmentation in general.

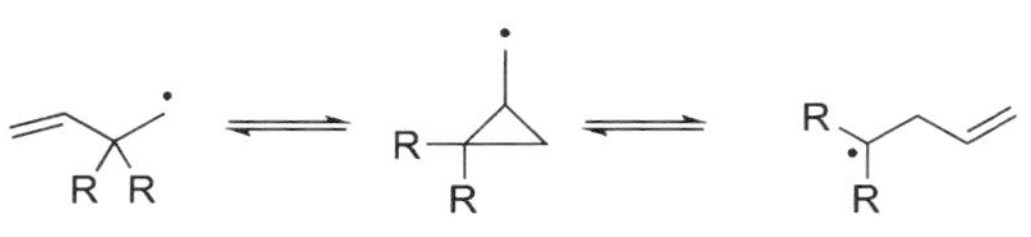

Scheme 2.12   Radical rearrangements in general.

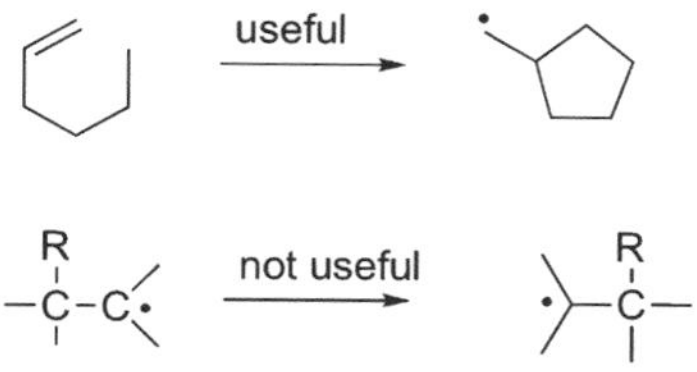

Scheme 2.13 Radical cyclization *versus* atom transfer reaction.

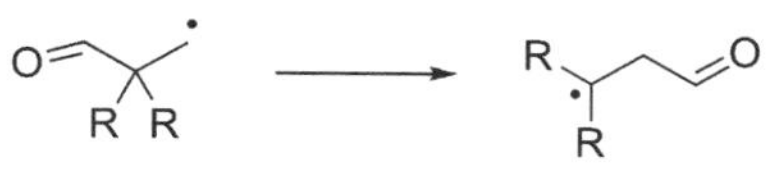

Scheme 2.14 Radical translocation reactions in general.

intermolecular reactions. The β-C–C bond of the intermediate cleaves rapidly because of the ring strain, which is the driving force for achieving the desired outcome in the transformation. The much smaller ring strain in the methyl-cyclopentyl radical intermediate stops the rearrangement of the hexenyl radical at the cyclopentylmethyl radical step (Scheme 2.13).

This is in accordance with the very slow 1,2-alkyl or -hydrogen shift in radicals, which does not occur in synthesis because of the strong σ-bond (Scheme 2.14).

2.16   Termination/Electron Transfer

The rate of electron transfer reactions depends on the difference in the redox potentials of educts and products. Since an alkyl radical possesses an unpaired electron in a non-bonding orbital, electron transfer reactions to many metals salts often occur with high rates. The higher are the SOMO energies of the radicals, the faster is the electron transfer.

2.17   Conclusion

With a brief historical background and general principles of free radical chemistry applicable to organic and aqueous media, we shall next turn our attention to the study of radical chemistry in aqueous media as its applies to organic synthesis and use the concepts that will emerge as the foundation upon which the rest of the material will rely and build upon.

References

1. M. Lehni, H. Schuh and H. Fischer, *Int. J. Chem. Kinet.*, 1979, **11**, 705; H. Langhals and H. Fischer, *Chem. Ber.*, 1978, **111**, 543.

2. H. R. Dutsch and H. Fischer, *Int. J. Chem. Kinet.*, 1981, **13**, 527; D. Griller and K. U. Ingold, *Acc. Chem. Res.*, 1976, **9**, 13; H. D. Beckhaus, G. Hellmann and C. Ruchardt, Chem. Ber., 1978, **111**, 72; K. Schluter and A. Berndt, Tetrahedron Lett., 1979, 929.
3. B. Giese, *Radicals in Organic Synthesis: Formation of Carbon–Carbon Bonds*, Pergamon Press, Oxford, 1986; B. Giese, *Angew. Chem. Int. Ed. Engl.*, 1985, **24**, 553.
4. A. L. J. Beckwith, Regio-selectivity and stereo-selectivity in radical reactions, *Tetrahedron*, 1981, **37**, 3073–3100; A. L. J. Beckwith, The pursuit of selectivity in radical reactions, *Chem. Soc. Rev.*, 1993, **22**, 143–151; B. Giese, Formation of CC bonds by addition of free radical to alkene, Angew. *Chem. Int. Ed. Engl.*, 1983, **22**, 753.
5. J. M. Tedder, The importance of polarity, bond strength and steric effects in determining the site of attack and the rate of free radical substitution in aliphatic compounds, *Tetrahedron*, 1982, **38**, 313–329.

CHAPTER 3

Why Water as a Solvent?
Reasons and Advantages

Until recently, the use of water as a solvent for organic reactions was mainly restricted to simple hydrolysis reactions.[1] Accordingly, most reagents and catalysts in organic synthesis have been developed for use in anhydrous, organic reaction media.[2] Why should we now spend time 'rediscovering' reactions for use in water that already work well in familiar organic solvents such as THF, toluene or methylene chloride? The answer is: because there are many potential advantages in replacing these and other unnatural solvents with water. The most obvious are the following:

- *Cost*: It does not get any cheaper than water!
- *Safety*: Most of the organic solvents used in the laboratory today are associated with risks: flammables, explosives, carcinogens, *etc.*
- *Environmental concerns*: The chemical industry is a major contributor to environmental pollution. With increasing regulatory pressure focusing on organic solvents, the development of non-hazardous alternatives is of great importance.

It is important, however, that the above-listed benefits are not gained at the expense of synthetic efficiency. Even a small decrease in yield, catalyst turnover or selectivity of a reaction can lead to a substantial increase in cost and the amount of waste generated. Fortunately, many theoretical and practical advantages of the use of water as a solvent for organic synthesis exist. These will be elaborated upon later but are briefly introduced here:

- Experimental procedures may be simplified since isolation of organic products and recycling of water-soluble catalysts and other reagents could be achieved by simple phase separation.
- Laborious protecting-group strategies for functionalities containing acidic hydrogens may be reduced.

RSC Green Chemistry No. 6
Radical Reactions in Aqueous Media
By V. Tamara Perchyonok
© V. Tamara Perchyonok 2010
Published by the Royal Society of Chemistry, www.rsc.org

- Water-soluble compounds could be used in their 'native' form without the need for hydrophobic derivatisation, again eliminating tedious protection–deprotection steps from the synthetic route.
- As will be amply exemplified in this review, the unique solvating properties of water have been shown to have beneficial effects on many types of organic reactions in terms of both rate and selectivity.

3.1   Solubility of Organic Compounds in Water

Most chemical reactions are performed in solvents. The solvent provides a reaction medium in which reactants can be mixed over a very wide concentration range. In general, a good solvent should readily dissolve all or most of the participating reactants, should not interact adversely with the reaction and should be easily separated during workup for facile isolation of products. On the basis of the knowledge of the chemical properties of the reactants, the chemist chooses a solvent that is suitable to meet these criteria. From this perspective, it is not surprising that water has found limited use as a solvent for organic reactions. In truth, the poor solubility of reactants and the deleterious effect on many organic transformations are the main obstacles to the use of water as a reaction solvent. Nonetheless, the fact that many of the most desirable target molecules, *e.g.* carbohydrates, peptides, nucleotides and their synthetic analogues, and also many alkaloids and important drugs are readily soluble in water is inconsistent with the disproportionate bias towards the use of organic solvents for their preparation. It can be argued that our shortcomings as synthetic chemists prompt the use of exhaustive protecting-group strategies, thus limiting the possibility of using water as solvent because of the low solubility of the reactants. Moreover, with the limited number of organic transformations in water that are currently available to the synthetic chemist, intermediates soluble in organic solvents are preferred to those soluble in water, even if it means adding extra synthetic steps for derivatisation. This may have particular relevance in carbohydrate chemistry.

Notwithstanding the above, many organic targets and their intermediates have very low solubility in water, which may lead to thwarting of reactions due to phase separation and inefficient mixing of reactants, although heterogeneous mixtures may retain at least partly the positive influence of water, sometimes with the aid of sonication or microwave heating. A variety of strategies have been investigated in order to extend the scope of water-based organic synthesis to embrace also highly hydrophobic reactants, and these will be briefly discussed below.[3]

3.2   Organic Cosolvents

One of the more efficient and versatile methods of increasing solubility and one that does not require modification of the solute is to use an organic cosolvent. The cosolvent reduces the hydrogen bond density of aqueous systems, so that it is less effective in squeezing out non-polar solutes from solution. Cosolvents can be structurally diverse, but they all carry hydrogen bond donor and/or acceptor groups for aqueous solubility and a small hydrocarbon region that serves to disrupt the strong hydrogen bond network of pure water, thereby increasing the solubility of non-polar reactants.[3] Some of the most commonly used cosolvents are the lower alcohols, DMF, acetone and acetonitrile. The increase in solubility, however, comes at the cost of many of the properties that make water a unique solvent for synthesis, such as high polarity, high cohesive energy density and the hydrophobic effect. The significance of this erosion of the bulk properties of water depends on the nature of the chemical reaction. Reactions that involve charged or highly polar species will suffer more from a decrease in solvent polarity than reactions involving only uncharged species. Likewise, reactions with a negative activation volume (Diels–Alder reaction, Claisen rearrangement) are expected to be adversely affected by the addition of a cosolvent because of the concurrent decrease in cohesive energy density. Nevertheless, because of the efficiency and flexibility of cosolvents in solubilising organic solutes, the major part of the development of reactions in aqueous media has been achieved with water–cosolvent mixtures.[4]

3.3 Ionic Derivatisation (pH Control)

Adding a positive or negative charge to an ionisable solute generally brings about a substantial increase in its solubility in water. Adjustment of the solution pH is therefore an efficient method of solubilising weak electrolytes in aqueous media.[5] This approach, of course, changes the chemical nature of the reactants and may limit its use as a method of solubilisation for synthetic purposes. For some types of reactions, however, the presence of a charged, highly polar moiety can have a very positive effect. For example, the reactions of diene-carboxylates, sulfonates or -ammonium salts with dienophiles in aqueous Diels–Alder reactions display significantly higher reaction rates than the corresponding neutral dienes.[6] It is generally accepted that the rate enhancement of the Diels–Alder reaction in water is at least partly due to the influence of the hydrophobic effect in this medium. In making the diene amphiphilic, increased solubility comes with the added bonus of an enhanced hydrophobic effect and faster reaction. This is similar to the effect that one would achieve by adding 'salting-out' agents known to increase the hydrophobic effect, only in this case the salt is the reactant itself.[7] When a buffered reaction is feasible, it is one of the more efficient ways to keep organic molecules solubilised. A potential practical advantage of using pH control in organic reactions is that the product may be recovered from solution by precipitation upon suitable adjustment of the pH or by extraction after addition of appropriate phase-transfer counterions.

3.4 Surfactants

An intriguing means of achieving aqueous solubility is by using surfactants. These are amphiphilic molecules, that is, they contain one distinctly polar and one distinctly non-polar region. In water, surfactants tend to orient themselves so that they minimize contacts between the non-polar region and the polar water molecules, and when the concentration of the surfactant monomer exceeds a certain critical value (critical micelle concentration), micellisation occurs. Micelles are spherical arrangements of surfactant monomers with a highly hydrophobic interior and a polar, water-exposed surface. Organic solutes interact with micelles according to their polarity: non-polar solutes are buried in the interior of the micelle, moderately polar molecules locate themselves closer to the polar surface and distinctly polar solutes will be found at the surface of the micelle. This compartmentalisation of solutes is believed to be responsible for the observed catalytic or inhibitory influence on organic reactions in micellar media.[8]

3.5 Hydrophilic Auxiliaries

Another method to increase the solubility of aqueous organic reactions is by grafting hydrophilic groups on to insoluble reactants. This strategy has been only cautiously explored for synthetic purposes but has a pivotal role in medicinal chemistry and modern drug design because of the low water solubility of many drugs, which causes limited bioavailability and thus reduced therapeutic efficacy. One way to improve the solubility of drugs is to convert them into water-soluble prodrugs through covalent attachment of a hydrophilic auxiliary. Ideally, the attachment should be of a transient and reversible nature, allowing for release of the parent drug from the auxiliary upon distribution, by either enzymatic or chemical means. The size and nature of the solubilising function range from small- to medium-sized, acidic and basic ionisable moieties (*e.g.* carboxylic acids).[9]

3.6 Conclusion

Aqueous radical chemistry predominates in biological processes and the development of synthetic radical aqueous chemistry will aid in our understanding of the detailed mechanism of the chemistry of life, such as biocatalysis, and in turn will have applications in evolving areas such as artificial biocatalysis, biomaterials and biotechnology in general. The main purpose of this book is to attempt to lift the greatest restriction on the implementation of aqueous radical synthesis, which is a misconception that might persist with many synthetic chemists regarding the inadequacy of aqueous radical chemistry as an equal partner in synthesis and chemistry in general.

References

1. C.-J. Li and T.-H. Chan, *Comprehensive Organic Reactions in Aqueous Media*, Wiley, New York, 2007.
2. D. C. Rideout and R. Breslow, *J. Am. Chem. Soc.*, 1980, **102**, 7816.

3. S. H. Yalkowsky, *Solubility and Solubilisation in Aquous Media*, Oxford University Press, New York, 1999.
4. U. Lindstrom, Stereoselective organic reactions in water, *Chem. Rev.*, 2002, **102**, 2751–2772.
5. B. D. Anderson and K. P. Flora, in *The Practice of Medicinal Chemistry*, P. H. Stahl and C. G. Wermuth (Eds.), Academic Press, London, 1996, 739–754.
6. P. A. Grieco, K. Yoshida and P. J. Garner, *Org. Chem.*, 1983, **48**, 3137; J. F. W. Keana, A. P. Guzikowski, C. Morat and J. J. Volwerk, *J. Org. Chem.*, 1983, **48**, 2661; P. A. Grieco, P. Galatsis and R. F. Spohn, *Tetrahedron*, 1986, **42**, 2847.
7. R. Breslow and C. J. Rizzo, *J. Am. Chem. Soc.*, 1991, **113**, 4340.
8. S. Tascioglu, *Tetrahedron*, 1996, **52**, 11113.
9. P. H. Stahl and C. G. Wemuth (Eds.), *The Practice of Medicinal Chemistry*, Academic Press, London, 1996.

CHAPTER 4

Reducing Agents Based on Group 4 and Aqueous Media

4.1 General Introduction

The use of free radical reactions in organic synthesis started with the reduction of functional groups. The purpose of this chapter is to give an overview of how the conventional tin-mediated reductions in organic solvents were transformed into more benign and metal-free reductions in aqueous media. The important players in this are Bu_3SnH, $(TMS)_3SiH$, mercury-based hydrogen donors, phosphorus-based hydrogen donors and thiols as efficient sources of facile hydrogen atom transfer agents for atom transfer. A number of reviews have discussed various aspects of these topics in detail.[1] The purpose of this chapter is to link the important aspects of various classes of hydrogen donors and produce a good reference point for future synthetic endeavours and investigation.

4.2 Mercury Hydrides and Water: Brief General Reaction Considerations

Of the other metals or metalloid hydrides that could be conceivably participate in radical processes, by far the most important from the synthetic point of view are organomercury hydrides. Interest in the free radical chemistry of organomercurials blossomed when it was established that the classical demercuration reaction with sodium borohydride was in fact a radical chain reaction (Scheme 4.1).[2]

Pasto and Gontarz found, for example, that both the *erythro* and *threo* isomers from the acetoxymercuration of *trans*- and *cis*-butene gave the same 50:50 mixture of *erythro*- and *threo*-3-deutero-2-butanol upon reduction with sodium borondeuteride (Scheme 4.2).[3] The complete loss of stereochemical information could only be rationalised by invoking a carbon radical intermediate.

RSC Green Chemistry No. 6
Radical Reactions in Aqueous Media
By V. Tamara Perchyonok
© V. Tamara Perchyonok 2010
Published by the Royal Society of Chemistry, www.rsc.org

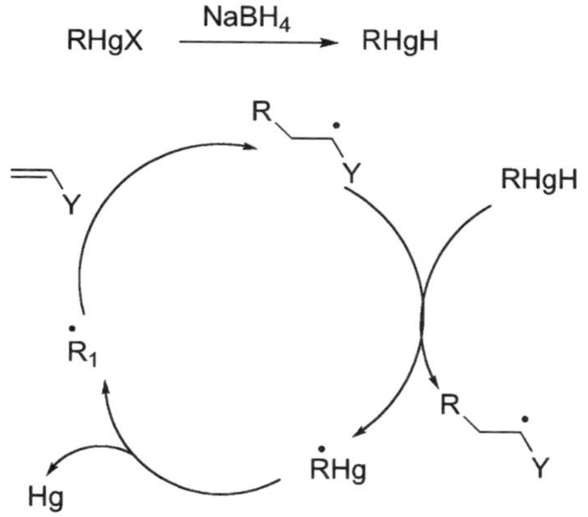

Scheme 4.1 Radical chain reduction using organomercury hydride.

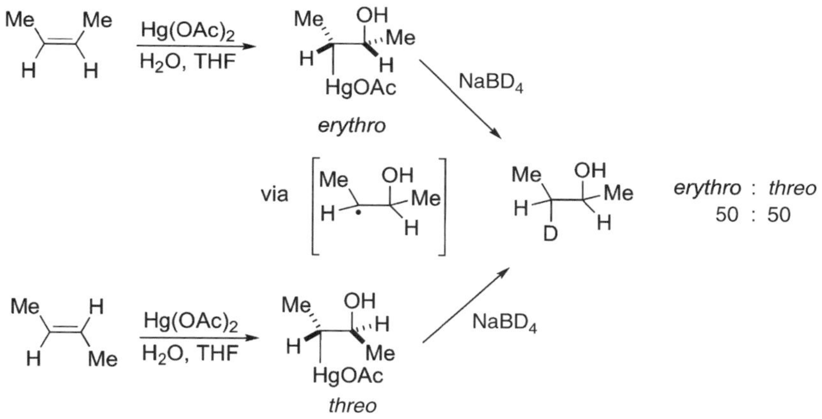

Scheme 4.2 Stereochemistry of the reductive demercuration reaction.

Further evidence was obtained from the observation that alcohols were obtained when reduction was performed under air.[4]

As described in Scheme 4.3, the influence of a labile organomercury hydride which gives rise to a carbon radical appears to be a logical explanation. Aliphatic organomercury hydrides are so labile that there is usually no need to add any initiator: traces of oxygen (or even some metallic impurities) are sufficient to trigger the radical chain.

The organomercury radical extrudes mercury to produce a carbon radical which propagates the chain by abstracting a hydrogen atom from another molecule of hydride (path A). If the reaction is conducted in the deliberate presence of oxygen, the carbon radical can be captured to give ultimately an alcohol by reduction of the intermediate hydroperoxide (path B).

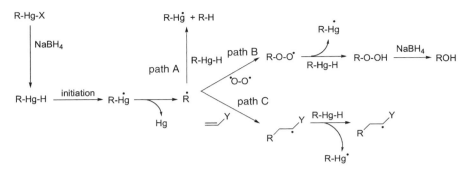

Scheme 4.3   Reactions mechanism of organomercury hydride.

The carbon radical may also be trapped in a variety of other ways such as with an external or internal alkene (path C). As with the tin- and silicon-based methods, the trapping processes has to compete with the hydrogen atom transfer step of path A. The rate of the latter has been estimated to be at least as high as $10^7\,M^{-1}\,s^{-1}$;[5] the trapping agent must therefore be especially reactive in order to succeed. In practice, the organomercury hydride is not isolated but simply generated *in situ* at room temperature or below by reduction of an organomercuric halide or carboxylate, usually with a borohydride, but other reducing agents, including stannanes, can also be employed. The rate of addition of the reductant controls the concentration of the hydride; metallic mercury separates out and is easily removed by decanting. A practical hint is that, especially for small-scale work, degassing of the solvent is especially important because of the relative solubility of oxygen in organic solvents at room temperature, unless alcohol is a desired product.

4.3   Application of Organomercury Hydrides in Synthesis

Organomercury hydrides constitute a convenient and synthetically highly useful source of carbon-centred radicals because of the diversity of methods allowing access to the precursors (Scheme 4.4). Oxymercuration of alkenes or cyclopropanes and mercuration of organometallic reagents are especially powerful and versatile in this regard. The mercuric group may be simply replaced by hydrogen or deuterium as in the mechanistic studies discussed earlier. Introducing deuterium or tritium *via* organomercury derivatives is especially easy.

This is shown by the first sequence in Scheme 4.4;[6] the organomercuric precursor in this case was constructed by a double alkoxymercuration reaction. Addition of sodium chloride caused the formation of the organomercuric chloride; this is usually insoluble and merely separated by filtration. An

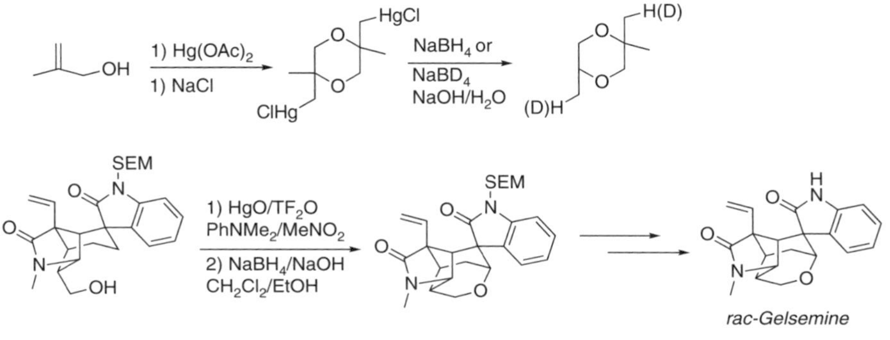

Scheme 4.4 Examples of oxymercuration/demercuration in organic and aqueous media.

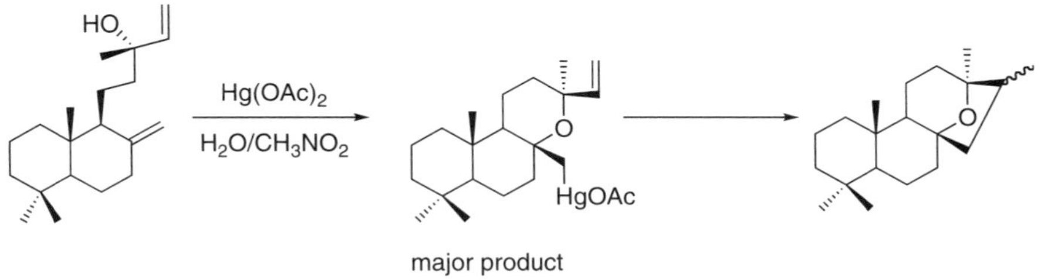

Scheme 4.5 Regioselective oxymercuration–demercuration.

intramolecular alkoxy-mercuration–demercuration sequence was used in the late stages of a synthesis of gelsemine.[7]

The concise construction of the strobane skeleton from 13-*epi*-manool (Scheme 4.5) provides an example of intramolecular capture of the intermediate carbon radical.[8] The acetoxymercuration step leads to two epimeric derivatives at C8 but only the β-isomer can undergo the desired cyclisation. This occurs fairly efficiently to give one isomer whose stereochemistry has not yet been established.

To conclude this part of the discussion, let us illustrate the synthetic potential of the mercuric acetate-induced opening of cyclopropane rings with the example shown in Scheme 4.6.[9] Acetoxymercuration of the trimethylsiloxycyclopropane in the absence of water gives a siloxyacetate which undergoes the normal demercurative addition to chloroacrylonitrile. When water is present, however, this compound is rapidly hydrolysed into the corresponding aldehyde and the intermediate radical can now rearrange with an efficiency that depends on the concentration of the olefinic trap: at higher dilution, the initial primary radical has time to interconvert into the more stable tertiary radical, which finally gives adduct **2**. Both isomeric products **1** and **2** can thus be prepared by a simple modification of the experimental conditions.

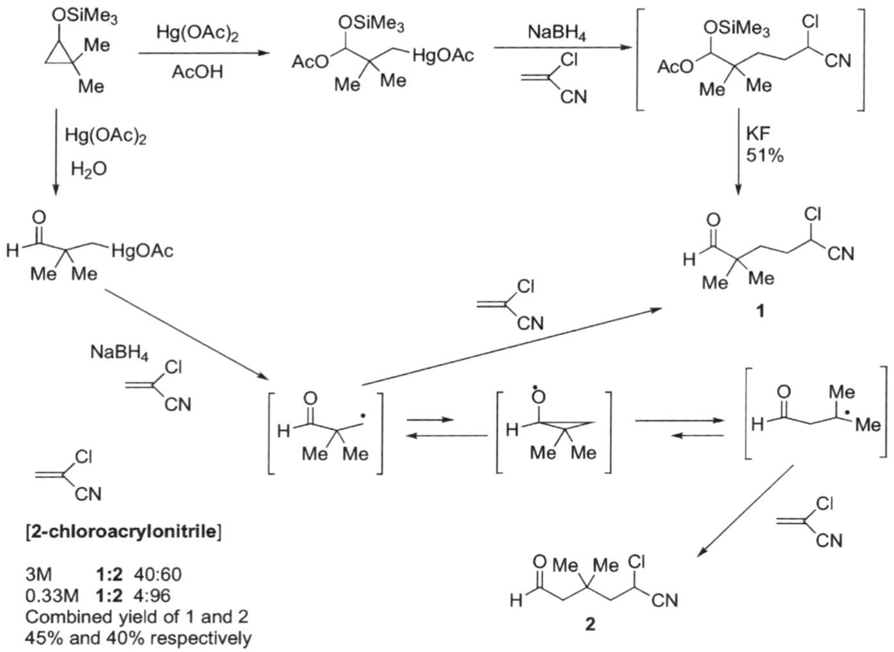

Scheme 4.6 Intermolecular addition reaction using the mercuration–demercuration reaction.

4.3.1 Perspective and Future Directions

As has been shown, mercury hydrides (RHgH), like tin hydride, can be used to generate alkyl radicals under mild conditions in aqueous and organic media. The mild reaction conditions and easy access to a wide variety of organomercury reagents has led to many synthetic applications. However, the main drawback of these reagents is the toxicity of organomercury compounds and the disposal of organomercury metal at the end of the reaction.

4.4 Bu₃SnH as a Major Player in Reductions in Organic Solvents: Would Water Be a Possible Solvent for the Transformations?

Chain reactions involving organotin hydrides are by far the most commonly employed in organic synthesis. A wide variety of carbon- and heteroatom-centred radicals can thus be generated and captured under conditions of sufficient mildness to be compatible with many of the highly complex intermediates encountered in modern synthesis. The propagation steps are generally fast enough that the steady-state concentration of the intermediate radical

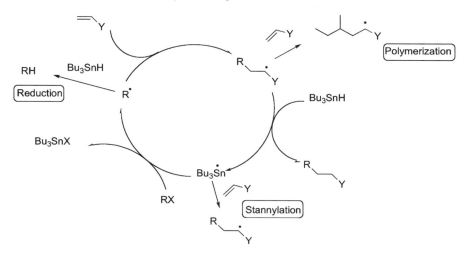

Scheme 4.7 Radical chain reactions using tri-*n*-butyltin hydride.

remains automatically low and unwanted radical–radical interactions do not occur. In general terms, the process can be written as shown in Scheme 4.7.

For simplification and clarity, tributyltin hydride is used in the scheme as it is the most common reagent, but the same will apply for other organotin hydrides, with perhaps some relatively minor variations in the reaction rates. However, before illustrating the synthetic potential of this process, its necessary to discuss in more detail the individual steps and some practical aspects of the transformation.

4.5 Radical Initiators: Choosing the Appropriate Radical Initiator for the Aqueous Free Radical Reaction of Choice

A survey of the literature reveals that there are several commercially available radical initiators used in free radical transformations which are conducted in both aqueous and ionic liquid media (see Figure 4.1 for structures). The most popular thermal radical initiator is $2',2'$-azobisisobutyronitrile (AIBN), which has a half-life ($t_{1/2}$) in toluene of 1 h at 81 °C and 10 h at 65 °C.[7,10]

Although AIBN is a popular choice, other azo compounds can give superior results due to their varying half-life. Indeed, the nature of the substitution of a newly formed carbon radical plays an important role in their half-life, as can be seen from the following for $2',2'$-azobis(4-methoxy)-3,4-dimethylvaleronitrile (AMVN), with a $t_{1/2}$ in toluene of 1 h at 56 °C and 10 h at 33 °C. There are also hydrophilic azo compounds, such as $2',2'$-azobis(2-methylpropionamidine) dihydrochloride (APPH), with a $t_{1/2}$ in water of 10 h at 56 °C. Typically, 5–10 mol% of the initiator is added, either in one portion or *via* slow addition. There

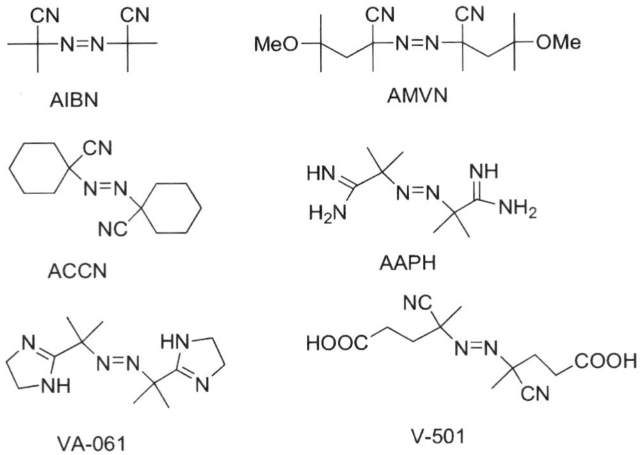

Figure 4.1 Commercially available azo free radical initiators.

are numerous advantages of the water-soluble azo initiators over organic peroxides, *e.g.* they are relatively inexpensive, do not produce undesirable decomposition products and/or toxic reaction by-products and they have added stability because of shock resistance. Peroxides such as *tert*-butyl per-benzoate [PhC(O)OOBu-*t*] and di-*tert*-butyl peroxide (*t*-BuOOBu-*t*), whose $t_{1/2}$ values are 1 h at 125 and 147 °C, respectively, are used from time to time.[4,8,10,11] Generally, peroxides are used when the reaction requires a more reactive initiating species and is most familiar to synthetic chemists.[4,8,10,11] They initially produces acyloxyl radicals, which often decarboxylate prior to undergoing bimolecular reactions, affording the equally reactive carbon-centred phenyl radical.

Photochemically generated radicals in chain reactions are less familiar to the synthetic chemist.[12] The above-mentioned peroxides have been used in the presence of light to initiate radical chain reactions at room or lower tempera-ture. Azo compounds are also known to decompose under such conditions and therefore have rarely been used under such conditions.

Triethylborane in the presence of a very small amount of oxygen is an excellent initiator for radical chain reactions. It has been known for a long time that trialkylboranes (R_3B) react spontaneously with molecular oxygen to give alkyl radicals ($R^\bullet$), but only in the last 10 years has this approach been suc-cessfully applied for the initiation of free radical reactions.[13] These reactions can be run at temperatures as low as −78 °C, which allows for a high degree of stereo-, regio- and enantioselectivity, which is often required in the synthesis of complex molecules. One practical way of conducting such reactions is to equip the reaction flask with a toy balloon filled with either nitrogen or argon and let oxygen from the atmosphere slowly diffuse into it and hence into the reaction mixture. It must be remembered that the rate of decomposition of the initiator determines the number of chains that are started and that the steady-state concentration of the intermediate radicals will depend on the efficiency of the propagation steps so that the experimental conditions (especially temperature, concentration and mode of addition of reagents) and the type and amount of initiator must be selected and eventually modified with these considerations in mind. Its is also advisable to avoid contact of the refluxing solvent with any rubber septa since additives in the rubber can leach into the reaction medium and act as inhibitors of the chain reaction.

Selection of the correct choice of the initiator is generally made according to the experimental conditions. For example, α-cyanoalkyl radicals derived from AIBN are capable of abstracting a hydrogen atom from (TMS)$_3$SiH, but not from Ph$_2$SiH$_2$. It is also advisable to add the initiators slowly during the course of the reaction, either in solution or portion-wise, with particular attention being paid to the half-life of the decomposition of the initiator at the operating temperature of the reaction. The direct addition at once of the thermal initiator together with a reaction temperature much higher than that for a 1 h half-life generally does not lead to the optimal outcome of free radical reactions. At low temperatures, when the thermal initiation is not viable, Et$_3$B–O$_2$ is the best starting point.

4.6 Reaction Rates and Practical Hints

The first propagation step concerns the reaction of the substrate R–X with the tributylstannyl radical. R–X can be a halide (but not a fluoride), a chalcogenide, an isocyanide, a nitro compound or some more complicated radical precursor group such as xanthates, Barton and/or Kim esters.[14] The rate of this step depends on X, the weakness of the R–X bond and the stability of the ensuing radical R$^•$ (roughly speaking, the more stable the radical the weaker is the R–X bond); some typical rate constants are collected in Table 4.1.[15]

As one would expect from the respective bond strengths, the order for the halides is R–I > R–Br ≫ R–Cl for the same R group and selenides react faster than sulfides for the chalcogenides. In general, when the bond strengths are comparable, monovalent halides tend to be more reactive than divalent

Table 4.1 Rates of the reaction of some carbon-centred radicals with organo-stannanes.[15]

Radical	Stannane	Rate constant $(M^{-1}s^{-1})^a$
CH$_3$$^•$	Bu$_3$SnH	$K_{30} = 1.2 \times 10^7$
RCH$_2$$^•$	Bu$_3$SnH	$K_{30} = 2.7 \times 10^6$
RCH$_2$$^•$	Ph$_3$SnH	$K_{25} = 5 \times 10^6$
Me$_2$CH$^•$	Bu$_3$SnH	$K_{30} = 1.5 \times 10^6$
Me$_3$C$^•$	Bu$_3$SnH	$K_{30} = 1.7 \times 10^6$
cyclo-C$_3$H$_5$$^•$	Bu$_3$SnH	$K_{30} = 8.5 \times 10^7$
Me$_2$C=CH$^•$	Bu$_3$SnH	$K_{30} = 3.5 \times 10^8$
C$_6$H$_5$$^•$	Bu$_3$SnH	$K_{30} = 5.9 \times 10^8$

^{a}Subscript numbers on K are temperatures in °C.

chalcogenides (*cf.* bromides and selenides), presumably for steric reasons. For the same X, aliphatic derivatives are much more reactive than vinylic or aromatic substrates and substituents such as a carbonyl group enhance the reactivity (*cf.* alkyl chlorides and α-chloro esters), reflecting the stability of the corresponding radicals. It can be seen that the absolute rate constants are high enough to belong to the ultra-fast domain but still sufficiently different to allow for a considerable amount of selectivity. Rates for such ultra-fast elementary radical processes are usually established by competing experiments, ultimately against each other, often unimolecular reactions, whose absolute rate has been determined directly by a sophisticated kinetic technique such as laser flash photolysis. These reference reactions provide a convenient scale and are termed radical clocks.[16]

Absolute rates for hydrogen abstraction form the stannane, the second propagation step, have also being measured.[17] These are indeed exceedingly efficient processes. These rates are on the higher end of the scale with relatively little spread (Figure 4.2). Aromatic, vinylic and cyclopropyl radicals react fastest, at close to the diffusion limit, whereas aliphatic radicals are some 100 times slower, with not much difference between primary, secondary and tertiary systems. Benzylic radicals, on the other hand, are the least reactive.

Tributyl- and triphenyltin hydrides are now commercially available, the former being much cheaper as their preparation involves merely distilling a mixture of hexabutylditin oxide and polymethylhydrosiloxane.[18]

The main practical difficulty when working with stannanes, apart from the somewhat nauseating stench, which seems to cling forever to glassware, is the purification of the product and the complete removal of organotin residues. Various tricks have been invented over the years, as follows:

- precipitation as organotin fluorides
- partitioning between methanol or acetonitrile and petroleum ether
- treatment with DBU
- treatment with sodium hydroxide
- use of molecular sieves.

In any case, its advisable to use freshly distilled material, as better yields are invariably obtained and the purification is often easier. In some instances, the

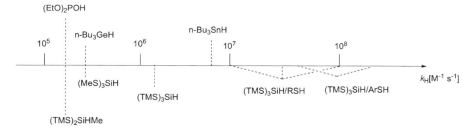

Figure 4.2   Rate constants for hydrogen atom transfer from a variety of hydrogen donor systems and primary radicals at 80 °C.

stannane can be used in a catalytic amount in conjunction with another reducing agent such as polymethylhydrosiloxane, sodium borohydride or cyanoborohydride. Polymer-supported reagents and stannanes with nitrogen or fluorous or polyaromatic substituents have also been proposed to simplify purification. Finally, organotin derivatives must be considered to be potentially toxic and appropriate care exercised while handling them, so there is a need for alternative hydrogen donors.

4.7 Tris(trimethylsilyl)silane: Historical Perspective and What Can Be Done 20 Years After its Discovery as a Radical-based Reducing Agent

Tris(trimethylsilyl)silane, $(Me_3Si)_3SiH$, was synthesised by Gilman and co-workers in 1965 and was then almost totally forgotten for the next 20 years,[19] until Chatgilialoglu and co-workers discovered that this material could serve as a radical-based reducing agent.[20] Twenty years of use of the $(Me_3Si)_3SiH$ reagent in various laboratories world-wide have revealed the potential of this reagent, thereby stimulating further interest in organosilanes as safe and efficient reagents for radical chemistry. Overall, organic free radical chemistry has suffered (and still does) from the use of toxic tin compounds. New reagents and reactions have been introduced and new modes of use for the reagents revealed, which will certainly motivate the abandonment of traditional choices. Additionally, the radical chemistry associated with $(Me_3Si)_3SiH$ has been applied to polymers and in materials science. For example, radical reactions have been found to be the most convenient method for the organic modification of hydrogen-terminated silicon surfaces. The radical chemistry of organosilanes has been the subject of periodical reviews, which mainly focused on $(Me_3Si)_3$-SiH.[20-22]

4.8 $(Me_3Si)_3SiH$ Reagent: How, When and Why to Use

The majority of radical reactions of interest to synthetic chemists are chain processes. In this context, $(Me_3Si)_3SiH$ is an effective reducing agent for the removal of a variety of functional groups. The need for non-toxic reagents to achieve this step is evident, since the compounds have to be tested for biological assays and even traces of toxic by-products can affect the results. From this simple example, it is clear that the arrival of the silane reagent on the scene resulted in a great advance for radical chemistry, since chemical and biological issues were satisfied simultaneously. Hydrogen abstraction from the sterically demanding $(Me_3Si)_3SiH$ preferably occurs from the less-hindered side of the intermediate cyclopropyl radical, thus affording a very high *cis* selectivity.[21]

The particular reactivity of $(Me_3Si)_3SiH$ has promoted further advances in fast trapping reducing systems. The $(Me_3Si)_3SiH$–thiol pair is a brilliant example of such systems.[22] The mechanism in Scheme 4.8 illustrates the propagation steps with 2-mercaptoethanol as the thiol. Its role is to act as a hydrogen donor and then to be regenerated by reaction of thiyl radical with silane. This reactivity can be employed to study the fast reaction of carbon-centred radicals with thiols, using the free-radical clock methodology. The fact that $(Me_3Si)_3SiH$ was developed with a strong contribution from physical organic chemistry in determining rate constants and thermodynamic parameters has certainly contributed to the confidence in its application due to the predictability of its behaviour.

Some physical and chemical characteristics of $(Me_3Si)_3SiH$ have also been important in its further development. $(Me_3Si)_3SiH$ is not soluble in water and does not suffer significantly after reaction with water at 100 °C for several hours, motivating interest in applying it to radical reactions in water. We trust that the survey presented here will serve as a platform to expand silicon radical chemistry with new and exciting discoveries. $(Me_3Si)_3SiH$ exemplifies the role of multidisciplinary approaches in developing new chemical tools and combining knowledge from physical organic chemistry with theoretical methodologies, organic synthesis and polymer and materials science.

4.9 Tris(trimethylsilyl)silane $(TMS)_3SiH$ as an Efficient Radical Hydrogen Donor, 'on Water' and 'in Water'

It was highlighted earlier that one of the drawbacks of conducting radical reactions in an aqueous medium is the immiscibility of either reagents, reactants or initiators under the reaction conditions. However, recently this viewpoint has changed and immiscibility has started to be considered advantageous.[23] Chemical reactions 'on water' are now given greater attention, since the reactivity in the suspension obtained by vigorous stirring of immiscible reactants seems to benefit from the enhanced contact surface of the resulting tiny drops, and also from the unique molecular properties developed at the interface between water and hydrophobic phase. 'Polarity reversal catalysis' seems to be ideal for free radical hydrogen transfer reactions 'on water'.[24] The advantages of employing this methodology are based on the fact that not only is the silane–thiol couple an efficient enhancer of the radical production and regeneration, but also the system is flexible enough to accommodate the amphiphilic thiol, in order to enhance the radical reactivity on the interface (Scheme 4.8).

4.10 On 'Polarity Reversal Catalysis' and the Benefits for Silicon-based Reducing Agents in Aqueous Media

The reaction of thiyl radicals with silicon hydrides [Reaction 1] is the key step of the so-called *polarity reversal catalysis* that has been utilised extensively in the radical chain reductions of alkyl halides and also in the hydrosilylation of

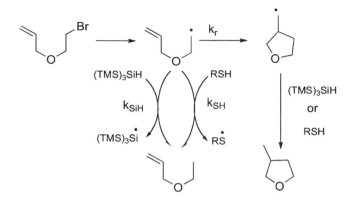

Scheme 4.8 Radical chain reduction/cyclisation of 3-(2-bromoethoxy)prop-1-ene using the (TMS)₃SiH–thiol (RSH) reducing system.

alkenes using the silane–thiol couple. The reaction is strongly endothermic and reversible.

$$RS^\bullet + R_3SiH \overset{k_{SH}}{\underset{k_{SiH}}{\rightleftharpoons}} RSH + R_3Si^\bullet \qquad \text{(Reaction 1)}$$

The rate constants k_{SH} and k_{SiH} were determined in cyclohexane at 60 °C relative to $2k_t$ for the self-termination of the thiyl radicals, using the kinetic analysis of the thiol-catalysed reduction of 1-bromooctane and 1-chlorooctane by silane, respectively.[24] The advantage of using a silicon hydride–thiol mixture lies in the low reactivity/solubility of alkyl- and/or phenyl-substituted organosilanes in reduction processes that can be ameliorated in the presence of a catalytic amount of alkanethiol.

Recently, Chatgilialoglu and co-workers reported that employing polarity reversal catalysis, *i.e.* the thiol–silane couple, represents not only an efficient synergy of radical production and regeneration, but also the use of an amphiphilic thiol, in order to enhance the radical reactivity at the interface. For the reduction of the organic halide (RX) by the couple (TMS)₃SiH–HOCH₂CH₂SH under radical conditions, the propagation steps suggested in Scheme 4.9 are expected. That is, the alkyl radical abstracts hydrogen from the thiol and the resulting thiyl radicals abstract hydrogen from the silane, so that the thiol is regenerated along with the chain carrying the silyl radical for a given RX (Scheme 4.9).[24]

Reduction of different organohalides, including bromonucleosides, was successfully carried out with yields ranging from 75% to quantitative, using (TMS)₃SiH in a heterogeneous system with water as the solvent. The procedure employed 2-mercaptoethanol as the catalyst and ACCN (1,1′-azobis(cyclohexanecarbonitrile) as organic solvent-soluble initiator or AAPH (2,2′-azobis (2-amidinopropane)dihydrochlorite) as water-soluble radical initiator,

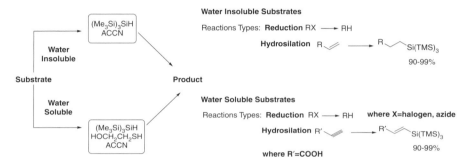

Scheme 4.9   Proposed mechanism of polarity reversal catalysis using the (TMS)$_3$SiH–thiol system.

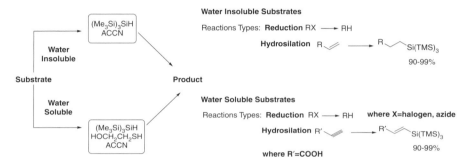

Scheme 4.10   Use of (TMS)$_3$SiH in aqueous media.

illustrating that (TMS)$_3$SiH can be used as the radical-based reducing agent of choice in an aqueous medium. (TMS)$_3$SiH does not suffer from any significant reaction with water and can safely be used with additional benefits, such as ease of purification and environmental compatibility.[24]

The extension and wide application of this work has come from the work of Chatgilialoglu and co-workers on the use of (Me$_3$Si)$_3$SiH as a hydrogen donor suitable for water-soluble and water-insoluble precursors, in systems comprising substrate, silane and initiator (ACCN) mixed at 100 °C. This system was proposed as it worked well for both hydrophilic and hydrophobic substrates, with the only variation that an amphiphilic thiol was also needed in the case of water-soluble compounds (Scheme 4.10).[25]

Recently, Chatgilialoglu *et al.* reported extensive studies on the use (TMS)$_3$SiH with classes of water-soluble and water-insoluble starting material in a broad range of free radical transformations such as HAT (Hydrogen Atom Transfer Reactions), hydrosilylation and radical cyclisation utilising (TMS)$_3$SiH under appropriate conditions.[25] The use of (TMS)$_3$SiH in water is attractive in view of its commercial availability and stability to aqueous, pH and oxidative conditions and also its optimal environmental compatibility. Two established methods have been utilised for the use of (TMS)$_3$SiH in water, depending on the hydrophilic or hydrophobic character of the substrates, as follows.[25]

4.10.1 Reduction of Water-insoluble Compounds

The products were obtained in excellent yields after simple hexane extraction. The removal of bromine and iodine proceeded smoothly. The removal of PhSe afforded methylcyclohexanone, indicating that the decarbonylation of acyl radicals takes place. The efficiency of deoxygenation of alcohols (Barton–McCombie reaction) is independent of the type of thiocarbonyl derivative (*i.e.* O-arylthiocarbonate, O-thioxocarbamate, thiocarbonylimidazole or xanthate), as previously reported for $(TMS)_3SiH$ in organic solvents.[25]

4.10.2 Reduction of Water-soluble Compounds

When the above-mentioned reaction conditions were used for the reduction of water-soluble compounds, very sluggish reaction outcomes were obtained, leading the authors to rely on the well-documented 'polarity reversal catalysis' idea such as utilising $(TMS)_3SiH$–SH as a hydrogen donor pair and achieving excellent reaction outcomes. The reduction of hydrophilic substrates in water has the added advantage of an easy separation of the silane by-products by partitioning between water and the organic phase. A few representative bromides and iodides were reduced in excellent yields, as reported in Scheme 4.11.

The reduction of an organic halide (RX) by the $(TMS)_3SiH$–RSH couple under radical conditions was proposed to occur through the propagation steps shown in Scheme 4.12, that is, the alkyl or aryl radical (R•) abstracts hydrogen from the thiol in the water phase and the resulting thiyl radicals migrate into a lipophilic dispersion of the silane and abstract a hydrogen atom, thus regenerating the thiol along with the chain carrying the silyl radical for a newly

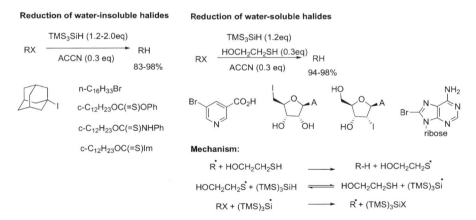

Scheme 4.11 Reduction of water-soluble and water-insoluble compounds in water using TMS_3SiH as an efficient hydrogen donor.

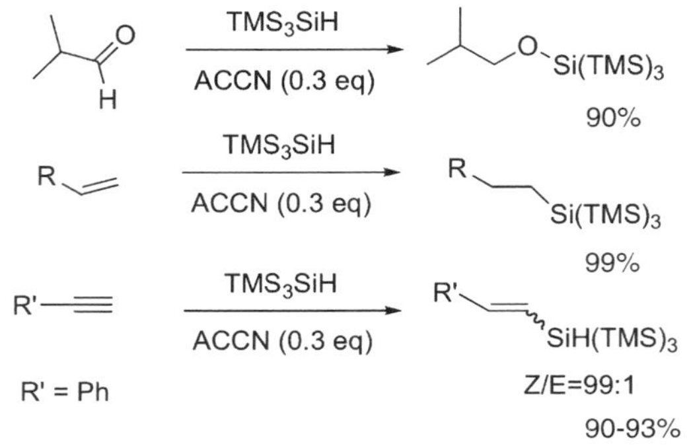

Scheme 4.12   Hydrosilylation of water-insoluble compounds in water.

formed RX. The reaction of the silyl radical with halide is expected to occur on the interface of the organic dispersion with the aqueous phase.

In the hydrosilylation of unsaturated bonds, using different hydrophobic compounds, such as aldehydes, alkenes and alkynes, under the same reaction conditions, the efficiency of the reaction was high and in all cases good to quantitative formation of the hydrosilylation products was achieved. Its also interesting to mention the higher *cis* stereoselectivity observed in water *versus* toluene ($Z:E = 51:49$ and 84:16 in toluene at 90 °C). It was concluded that the nature of the reaction medium does not play an important role either in influencing the efficiency of the radical transformation or in the ability to dissolve the reagent. It can be speculated that all water-insoluble materials (substrates, reagents and initiators) suspended in an aqueous medium can interact, due to the vigorous stirring that creates an efficient vortex and dispersion. Its was also proposed that radical initiation benefits from the enhanced contact surface of tiny drops containing (TMS)$_3$SiH and radical initiator.

The hydrosilylation of water-soluble propiolic acid was also tested under the same conditions. The reaction proceeded efficiently, giving the (*Z*)-alkene in optimal yield. It is worth mentioning that this hydrosilylation has recently been reported to be efficient also in neat conditions, where the initiation was linked to the presence of adventitious oxygen.[24,26]

Another successful class of radical reductions in water has been obtained by the transformation of azides to primary amines under the previously described experimental conditions (Scheme 4.13).[24]

Again, no reaction was observed in the absence of mercaptoethanol. The mechanistic steps for the transformation were elucidated by the Chatgilialoglu *et al.* and are represented in Scheme 4.13, in analogy with the pathway reported for the radical reduction of aromatic azides with triethylsilane in toluene, with the addition of silyl radicals to the azide function, liberation of nitrogen and formation of silyl-substituted aminyl radical. The thiol is the hydrogen atom donor to this intermediate and it can be regenerated by its interaction with the silane, thus propagating the chain. The hydrolysis of the silylamine occurred

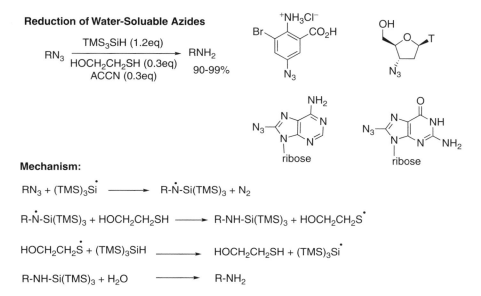

Reduction of Water-Soluable Azides

$$RN_3 \xrightarrow[\substack{HOCH_2CH_2SH\ (0.3eq) \\ ACCN\ (0.3eq)}]{TMS_3SiH\ (1.2eq)} RNH_2 \quad 90\text{-}99\%$$

Mechanism:

$$RN_3 + (TMS)_3\dot{Si} \longrightarrow R\text{-}\dot{N}\text{-}Si(TMS)_3 + N_2$$

$$R\text{-}\dot{N}\text{-}Si(TMS)_3 + HOCH_2CH_2SH \longrightarrow R\text{-}NH\text{-}Si(TMS)_3 + HOCH_2CH_2\dot{S}$$

$$HOCH_2CH_2\dot{S} + (TMS)_3SiH \longrightarrow HOCH_2CH_2SH + (TMS)_3\dot{Si}$$

$$R\text{-}NH\text{-}Si(TMS)_3 + H_2O \longrightarrow R\text{-}NH_2$$

Scheme 4.13   Reduction of water-soluble azides in water and reaction mechanism for the transformation in question.

rapidly in water and afforded the amine as a final product. In summary, the use and flexibility of the $(TMS)_3SiH$ as a versatile radical reagent in water have been cemented but not yet fully exploited.

References

1. C.-J. Li, Organic reactions in aqueous media with a focus on C–C Bond formation: a decade update, *Chem. Rev.*, 2005, **105**, 3095; R. A. Sheldon, Green solvents for sustainable organic synthesis: state of the art, *Green Chem.*, 2005, **7**, 267; C.-J. Li and L. Chen. Organic chemistry in water, *Chem. Soc. Rev.*, 2006, **35**, 68; V. T. Perchyonok, I. N. Lykakis and K. L. Tuck, Recent advances in C–H bond formation in aqueous media: a mechanistic perspective, *Green Chem.*, 2008, **10**, 153–163; V. T. Perchyonok, I. N. Lykakis, Recent advances in free radical chemistry of C–C bond formation in aqueous media: from mechanistic origins to applications, Mini-Rev. *Org. Chem.*, 2008, **5**, 19–32.
2. G. M. Whitesides and J. SanFillipo, *J. Am. Chem. Soc.*, 1970, **92**, 6611; C. L. Hill and G. M. Whitesides, *J. Am. Chem. Soc.*, 1974, **96**, 870; M. Devaut, *J. Organomet. Chem.*, 1984, **220**, c27.
3. D. J. Pasto and J. A. Gontarz, *J. Am. Chem. Soc.*, 1969, **91**, 719.
4. C. Walling, *Tetrahedron*, 1985, **19**, 3887–3900.
5. B. Giese and G. Kretzschmar, *Chem. Ber.*, 1984, 3160.

6. C. M. Hill and G. M. Whitesides, *J. Am. Chem. Soc.*1974, **96**, 870.
7. Y. Kita and M. Matsugi, in *Radical in Organic Synthesis*, eds. P. Renaud and M. P. Sibi, Wiley-VCH, Weinheim, 2001, **vol. 1**, pp. 1–10.
8. K. Itami and J. Yoshida, *Chem. Rec.*, 2002, **2**, 213–224.
9. C. Chatgilialoglu, *Helv. Chim. Acta.,* 2006, **89**, 2387–2398, and references cited therein; C. Chatgilialoglu, Organosilanes in *Radical Chemistry*, Wiley, Chichester, 2004, and references cited therein.
10. D. J. Adams, P. J. Dyson and S. J. Taverner, *Chemistry in Alternative Reaction Media*, Wiley-Interscience, Weinheim, Germany, 2004.
11. W. M. Nelson, *Green Solvents for Chemistry, Perspectives and Practice*, Oxford University Press, New York, 2003.
12. H. Yorimitsu, T. Nakamura, H. Shinokubo, K. Oshima, K. Omoto and H. Fujimoto, *J. Am. Chem. Soc.*, 2000, **122**, 11041–11047.
13. C. Galli and T. Pau, *Tetrahedron*, 1998, **54**, 2893–2904.
14. D. P. Curran, C. P. Jasperse and M. J. Totleben, *J. Org. Chem.*, 1991, **56**, 7169–7162.
15. D. Griller and K. U. Ingold, *Acc. Chem. Res.*, 1980, **13**, 317–323.
16. M. Newcomb, *Tetrahedron*, 1993, **49**, 1151–1176.
17. L. J. Johnston, J. Lusztyk, D. D. M. Wayner, A. N. Abeywickreyma, A. L. J. Beckwith, J. Scaiano and K. U. Ingold, *J. Am. Chem. Soc.*, 1985, **107**, 4594–4596.
18. K. Hayashi, J. Iyoda and I. Shiihara, *J. Organomet. Chem.*, 1967, **10**, 81–94.
19. H. Gilman, W. H. Atwell, P. K. Sen and C. L. Smith, *J. Organomet. Chem.*, 1965, **4**, 163.
20. C. Chatgilialoglu, *Acc. Chem. Res.*, 1992, **25**, 188.
21. C. Chatgilialoglu, C. Ferreri and T. Gimisis, in *The Chemistry of Organic Silicon Compounds*, Vol. 2, Part 2 (eds. Z. Rappoport and Y. Apeloig), Wiley, Chichester, 1998, pp. 1539–1579.
22. B. P. Roberts, *Chem. Soc. Rev.*, 1999, **28**, 25, and references cited therein.
23. A. Chanda and V. V. Fokin, Organic synthesis 'on water', *Chem. Rev.*, 2009, **109**, 725–748, and references cited therein.
24. A. Postigo, C. Ferreri, M. L. Navacchia and C. Chatgilialoglu, *Synlett*, 2005, **18**, 2854–2856.
25. C. Chatgilialoglu, A. Postigo, C. Ferreri and S. Kopsov, *Org. Lett.*, 2007, **9**, 5159.
26. K. Miura, K. Oshima and K. Utimoto, *Bull. Chem. Soc. Jpn.*, 1993, **66**, 2356.

CHAPTER 5

On the Use of Phosphorus Hydrides as Efficient Hydrogen Donors

The organic chemistry of phosphorus is based on the rich array of stable compounds featuring a carbon–phosphorus bond. As a consequence, reactions involving organophosphorus radicals have a long and well-written history.[1] The last few years have witnessed a renewed awareness that P-centred radicals, especially those containing P–O bonds, could be of practical synthetic utility.[2] In this chapter, the first part briefly lays out the physical organic background of such intermediates. In the second part, the use of organophosphorus radicals possessing a P–H bond that can undergo homolytic cleavage as an alternative mediator is detailed. The third part is focused on radical additions of phosphorus-centred radicals to unsaturated compounds, an old reaction with a green flavour. Radical eliminations of phosphorus-centred radical are introduced in the fourth part.

5.1 Introduction: Hypophosphorous Acid (H₃PO₂) as a Free Radical Hydrogen Donor in Aqueous Media

The physical aspects underlying the reactivity of P-centred radicals are essential in order to benefit from the full potential offered by phosphorus-containing compounds. Hydrogen atom transfers (HATs), halogen atom abstractions, cyclisations and additions are the 'tool box' of the widely applied free radical chemistry to date. Several authors have determined key rate constants through various physical methods, including time-resolved ESR spectroscopy and flash photolysis.[3]

It is interesting that hydrogen abstraction to form phosphonyl radicals is comparable to that of phenyl-substituted silanes $(1.2 \times 10^5 \ M^{-1}s^{-1})$.[4] On the other hand, Turro and Jockusch showed that phosphinoyl radicals [R₂P(O)·] are

RSC Green Chemistry No. 6
Radical Reaction in Aqueous Media
By V. Tamara Perchyonok
© V. Tamara Perchyonok 2010
Published by the Royal Society of Chemistry, www.rsc.org

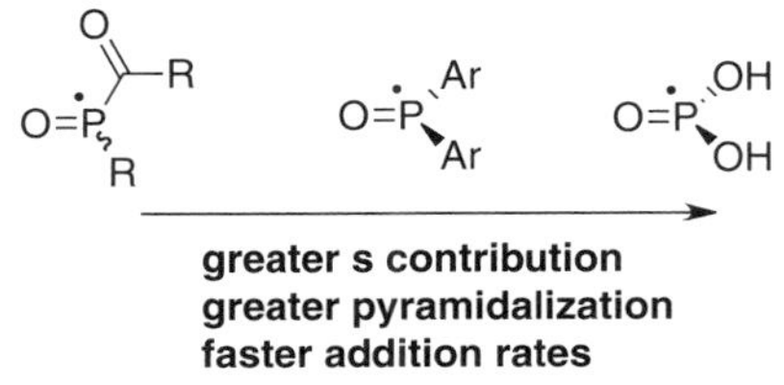

greater s contribution
greater pyramidalization
faster addition rates

Figure 5.1 Geometry of the P-centred radical.

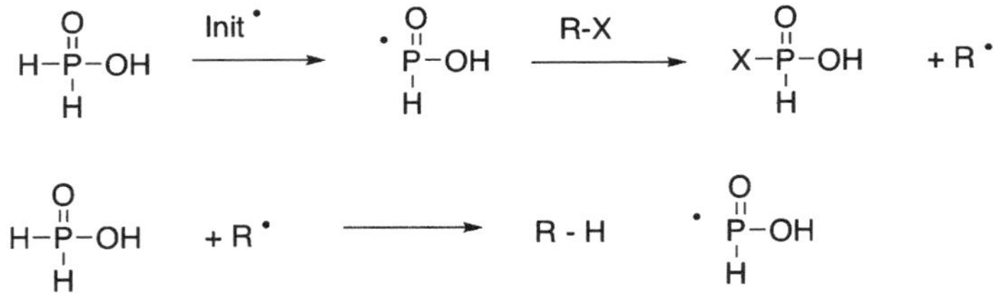

Scheme 5.1 Mechanism of HAT with phosphorus-based H-donor.

roughly 10 times more prone to reduction by thiophenol than are acylphosphinoyl radicals $[(RCO)_2P(O)^•]$.[4,5] A considerable amount of work has linked the structures and reactivities of the radicals. It had been shown that phosphinoyl and phosphonyl radicals are non-planar and as a result had a variable degree of s-character.[6] Phosphonyl radicals are more bent than phosphinoyl radicals. Acylphosphinoyl radicals are further flattened (Figure 5.1). [7] In general, the more bent the radical, the faster it adds to olefins. This trend has been attributed to the relative accessibility of the localised spin to the trapping agent.[3,8]

5.2 On the Use of P-centred Radicals as Efficient Hydrogen Donors

Radical chemistry has relied on the use of tributyltin hydride (TBTH) as a mediator.[2] Although extremely useful and robust, this compound is very toxic and its by-products are difficult to remove from reaction mixtures. For this reason and because radical reactivities are complementary to other reactivities, the quest for alternative mediators/hydrogen donors is extremely active. Thus, recent years, hypophosphorus acid and its derivatives have been used as mediators in several radical reactions in aqueous media. The mechanism of action of the P–H-containing phosphorus compounds is shown in Scheme 5.1.[2]

The blossoming area of the suitability of the hypophosphorous acid and its derivatives as hydrogen donors comes from the crucial re-examination by Barton and Jaszberenyi's group of some very old radical reductions involving hypophosphorous acid.[9] These workers' seminal contribution to the use of water as a solvent for water-soluble organohalides to be reduced cleanly and efficiently initiated and propagated the growth of the novel area of green and environmentally friendly methodology and application in free radical synthesis. This work has since been expanded by several groups. Several classes of novel hydrogen donors were introduced by Jang and co-workers, which expanded the 'reagent toolbox' of phosphorus-based hydrogen donors, such as sodium salts of hypophosphorous acid that can reduce water-soluble organohalides in water and the introduction of dibutyl- and diphenylphosphine oxides as novel, less hygroscopic reducing agents to accommodate the hygroscopic and hydrophilic substrates and reagents suitable for transformations of interest.[10,11]

The groups of Murphy and Stoodley both reported that *N*-ethylpiperidine hyposphosphite (EPHP) could trigger the formation of carbon–carbon bonds either through a 5-*exo-trig* cyclisation of an aryl radical obtained from an iodide or through a 5-*exo-dig* cyclisation of an alkyl α-keto radical obtained from a corresponding bromide; in both cases, the yield of the densely functionalised products was fairly high (Scheme 5.2).[12]

Kita and co-workers built on the results of previous studies to report EPHP-mediated cyclisation of hydrophobic substrates in water.[13] This breakthrough was made possible by running the reactions in the presence of a water-soluble initiator (VA-061) and a surfactant [cetyltrimethylammonium bromide (CTAB)]. The authors explained this outstanding result as being due to the micellar effect generated by CTAB, as in its absence the reaction proceeds to the extent of only 25%. The organic ammonium group of CTAB probably contributes to the incorporation of the hypophosphoric acid in the micelles (Scheme 5.3).

Murphy and co-workers introduced the use of a water-soluble phosphine oxide to synthesise indolones in excellent yields in water by a radical reaction array (aryl radical formation, HAT, cyclisation and rearomatisation) mediated by the reagent diethylphosphine oxide (DEPO).[15] The reaction also features yet

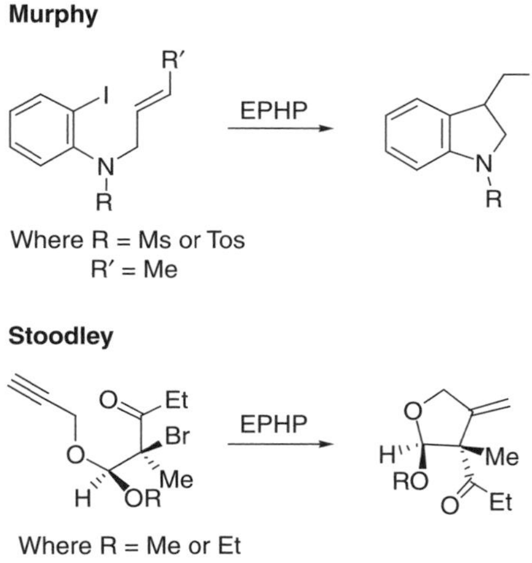

Murphy

Where R = Ms or Tos
R' = Me

Stoodley

Where R = Me or Et

Scheme 5.2 Cyclisations mediated by EPHP.

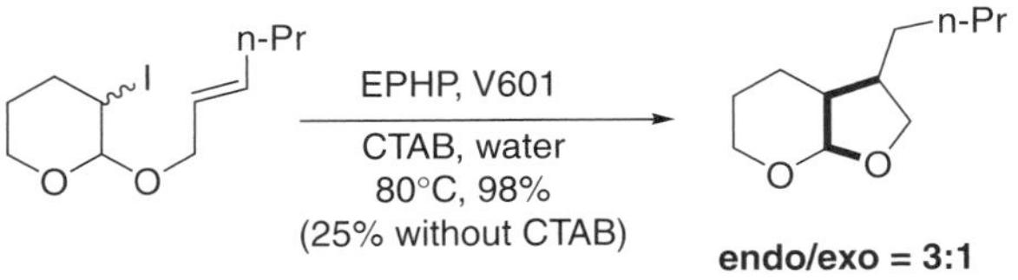

Scheme 5.3 Radical cyclisation in water.

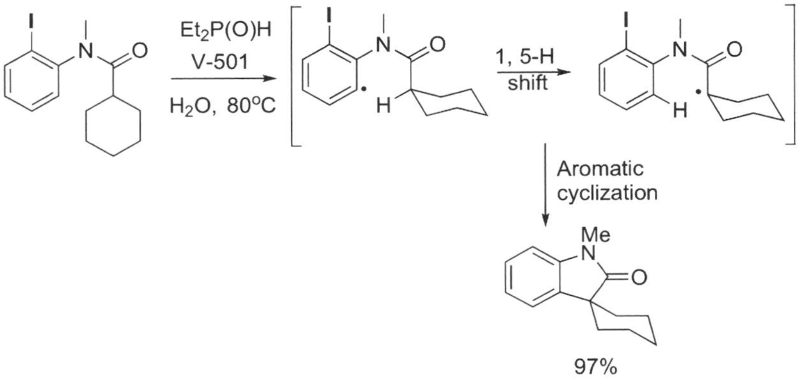

Scheme 5.4 DEPO-mediated arylation of lactams in water.

another type of water-soluble radical initiator, such as V601; with no other additives necessary. The remarkable advantage of these reaction conditions is that the reaction temperatures are significantly lower than are typically required for an efficient reaction with toxic TBTH in benzene. This permits significantly higher isolated yields than in the corresponding reaction mediated by ethylpiperidine hypophosphite (EPHP). Because DEPO is more lipophilic than hypophosphorous acid yet still water soluble, it can facilitate the interaction between the water-soluble mediator and initiator, and the initiator and the lipophilic substrates, without requiring a phase-transfer agent.[16] Also of importance is that the pK_a of DEPO is 6, which ensures that the excess of this almost neutral reagent can be extracted into the basic layer during workup (Scheme 5.4).

5.3 Deoxygenation Reactions in Aqueous Media

Tetraalkylammonium hypophosphites (TAHPs), as reported by Jang and co-workers, were prepared by mixing the tetraalkylammonium hydroxide with aqueous H_3PO_2, both of which are mild and efficient reagents for the radical deoxygenation of alcohols and formation of carbon–carbon bonds in water without the need for additives such as surfactants.[10,14] Cyclododecyl *S*-methyldithiocarbonate has been utilised as a model compound. The reaction of the xanthate with TAHP-1 in the presence of a water-soluble radical initiator, V-501 [4,4'-azobis(4-cyanovaleric acid)], in water yields 94% of the deoxygenated product after 5 h. Several other tetraalkylammonium hypophosphites were also used in the study and the results are summarised in Table 5.1.

This methodology was applied to synthesis of 2',3'-didehydro-2',3'-dideoxynucleosides, potent anti-HIV agents. The bis-xanthate of the N^3-methyluridine derivative was subjected to radical reaction conditions in water, giving the corresponding alkene in 82% yield. Similar results were also achieved in the case of the adenosine derivative (76% yield). However, Kita *et al.* found that the combination of a water-soluble radical initiator, 2,2'-azobis[2-(2-imidazolin-2-yl)propane] (VA-061), a water soluble chain carrier, 1-ethylpiperidine hypophosphite (EPHP), and a surfactant, cetyltrimethylammonium bromide (CTAB), resulted in a radical cyclisation that occurred in water with a variety of hydrophobic substrates[16] (Table 5.2).

Table 5.1 Deoxygenation of *O*-cyclododecyl *S*-methyl dithiocarbonate with tetraalkylammonium hypophosphites in water.

Entry	TAHP (equiv.)	V-501 (equiv.)	Time (h)	Yield (%)
1	TAHP-1 (3)	0.5	5	94
2	TAHP-1 (2)	0.5	8	84
3	TAHP-2 (3)	0.75	7	90
4	TAHP-3 (3)	0.75	8	62 $(29)^a$
5	TAHP-4 (3)	0.75	8	45 $(52)^a$

^{a}Recovered starting material.

Table 5.2 Radical cyclisation reactions using various initiators, hydrogen donors and surfactants.

highly hydrophobic

Initiator EPHP	Yield (%) (cis:trans)	H-donor/V-061	Yield (%) (cis:trans)	Additive EPHP V-061	Yield (%) (cis:trans)
VA-061	98 (55:45)	None	None	None	64 (74:26)
AIBN	19 (57:43)	(TMS)$_3$SiH	94% (67:33)	CTAB	98 (55:45)
Et$_3$B	50 (67:43)	EPHP	98% (55:45)	SDS	98 (51:49)
V-501	72 (51:49)	H$_3$PO$_2^-$ NaHCO$_3$	84% (78:22)	Triton X-100	98 (62:38)
VA-044	95 (50:50)	NaH$_2$PO$_2$	58% (78:22)	Et$_4$N$^+$Br$^-$	85 (65:35)

Table 5.3 Various free radical reactions at room temperature in the presence of chiral and achiral tetrasubstituted-ammonium hypophosphites in water using 0.5 equiv of Et_3B–air as radical initiator.

Entry	QAHP	Radical precursor	Product	Yield (%)
1	$(Bu_4N)^+H_2PO_2^-$			65
2	$(Bu_4N)^+H_2PO_2^-$			64
3	$(Bu_4N)^+H_2PO_2^-$			73
4	$(Bu_4N)^+H_2PO_2^-$			87
5	$(Bu_4N)^+H_2PO_2^-$			47
6	$(Bu_4N)^+H_2PO_2^-$			86

77

82

76

7 $(Bu_4N)^+ H_2PO_2^-$

8 $(S,S)\text{-}HMe_2NCHMePh^+ H_2PO_2^-$

9 $(S,S)\text{-}HMe_2NCHMePh^+ H_2PO_2^-$

Although the yields were not as high in the case of (TMS)₃SiH in comparison with the optimised EPHP–surfactant–initiator combination, the authors showed that tristrimethylsilylsilane [(TMS)₃SiH)] could be used as a silicon-based hydrogen donor in water/aqueous medium for the first time. The importance of this observation opened the door for the application of (TMS)₃SiH as an environmentally benign and powerful hydrogen donor in organic solvents in HAT 'on water' and 'in water'.

Recently, Perchyonok *et al.* reported further developments and applications of a broad range of fundamental free radical reactions, such as HAT, radical deoxygenations and radical cyclisations utilising quaternary ammonium salts of hypophosphorous acids as chiral and achiral hydrogen donors at room temperature, and the results are summarised in Table 5.3.[17] The same reactions were repeated at 80 °C under predetermined reaction conditions using either AIBN (10 wt%) (an organic-soluble radical initiator) or V-501 (10 wt%) (a water-soluble radical initiator) and comparable results were obtained. The results exceeded expectations as not only did the reactions proceed with good to excellent yields but they also showed a degree of stereoselectivity and enantioselectivity. The results represent the first examples of enantioselectivity being observed in free radical hydrogen transfer reactions in aqueous media and work is currently in progress to explore this novel aspect of enantioselectivity in an aqueous environment. The advantages of these hypophosphite reagents (chiral and achiral hypophosphinates) lie in their affordability, low toxicity, avoidance of the use of highly toxic 'tin-based hydrogen donors' and green reaction conditions.

The work was also extended to the application of functionalised emulsion-templated polymers (polyHIPEs) with CTAB as a suitable medium and also a source of *in situ* release of the surfactant for a wide range of fundamental free radical reactions.[18]

5.4 Applications in Synthesis

Hypophosphite hydrides can add to unsaturated compounds and thus lead to the formation of the corresponding phosphinates. Reding and Fukuyama designed a highly selective and elegant synthesis of indoles by reacting hypophosphite salts with unsaturated thioanilides.[19] Initial regioselective addition of the P-centred radical on to the C=S bond generated a new stabilised carbon radical that could cyclise on to the double bond in the *ortho* position, thus giving rise to the skeleton of indoles. Aromatisation of the compound generated the desired 2,3-substituted indoles. The authors employed this reaction as a key step towards the total synthesis of (±)-catharantine, a presumed biological precursor of the antitumour alkaloids vinblastine and vincristine (Scheme 5.5).

One of the most impressive synthetic achievements with the P-based mediators is the deoxygenation of an erythromycin B derivative towards the industrial synthesis of ABT-229, a potent motilin receptor agonist. Clean

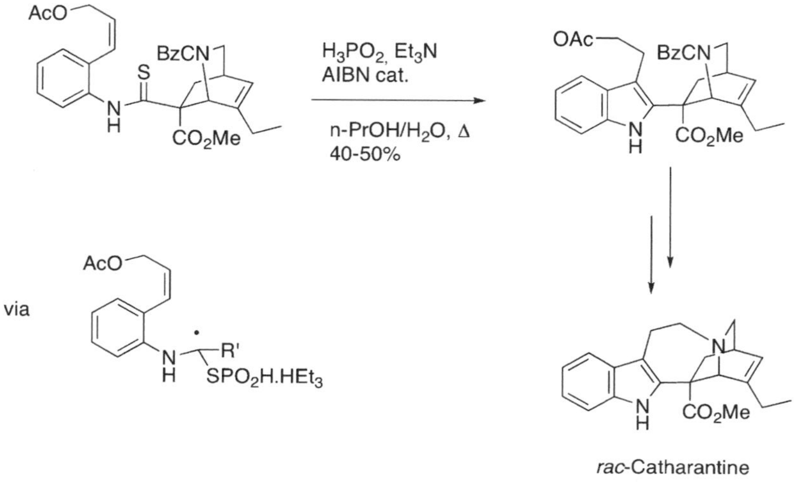

Scheme 5.5 Formation of indoles from thioanilides.

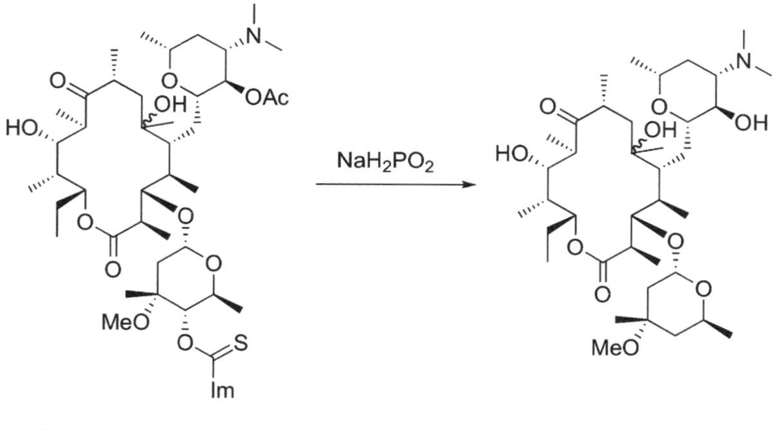

Scheme 5.6 Use of phosphinic acid in synthesis.

deoxygenation was ahieved on a 15 kg scale by using NaH_2PO_2 in an aqueous alcohol and phase-transfer conditions (Scheme 5.6).[20]

5.5 Deuterium Labelling and Radicals

Recent progress in the fields of NMR spectroscopy and mass spectrometry has allowed stable isotope labelling to become an important technique in metabolic research for determining the biological behaviour of small molecules. Stable isotopomers are more easily prepared and handled than their radioisotopic analogues. The development of synthetic methods for the preparation of compounds labelled with non-radioactive isotopes such as deuterium has therefore gained importance. Common methods for incorporating deuterium into organic molecules include ionic reactions using deuterides such as $NaBD_4$ and $LiAlD_4$, the triethylsilane–boron trifluoride etherate system ($Et_3SiD–BF_3$) and radical methods involving n-Bu_3SnD. However, radical reactions using tin deuterides have serious drawbacks when used in the synthesis of biologically active compounds, because the inherent toxicity of organotin derivatives and the difficulty of removal of residual tin compounds often prove fatal. Recently reported radical cyclisations have employed non-toxic and easily removed hypophosphorous acid (H_3PO_2) in aqueous solvents. Oshima and co-workers introduced a broad range of phosphorus-based deuterating agents, such as deuterated hypophosphorus acid potassium salts, deuterated phosphinic acid–DBU mixture and deuterated hypophosphorous acid–DCl–K_2CO_3 combination, to achieve radical deuteration.[21]

Deuteration of hydrophobic substrates was possible, albeit the incorporation of deuterium was not optimal, whereas less hydrophobic substrates led to deuteration with total incorporation (Scheme 5.7).[21] It is noteworthy that in

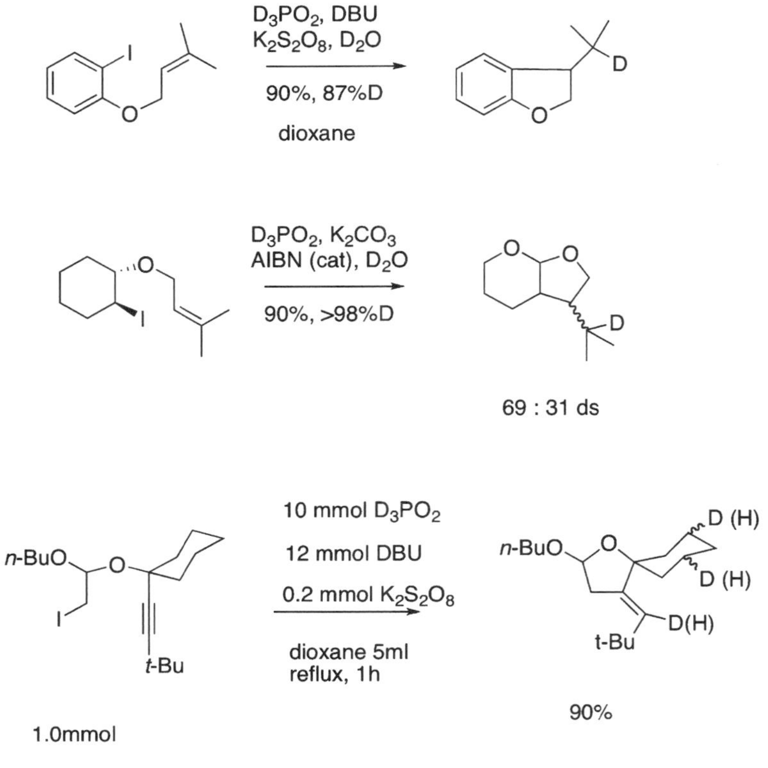

Scheme 5.7 D_3PO_2-mediated radical reactions in aqueous medium.

this radical reaction a quaternary centre bearing a deuterium is created in high yield. Various hydrogen–deuterium exchanges were conducted in either D_2O or D_2O–dioxane mixture and complete deuterium incorporation could be achieved by the reactions in D_2O without an organic co-solvent and an organic base.

5.6 Conclusion

The growing awareness of green and sustainable methods stemming from legitimate concerns for the environment gives a strong and renewed relevance to the development of radical reactions involving phosphorus. Work in this field has moved from the understanding of physical aspects underlying the reactivities to optimisation of the latter and their use in the total synthesis of complex molecules. Further progress should come from the introduction of more atom-efficient and stereoselective methods. Given the existing knowledge concerning phosphorus-containing compounds, progress will certainly also be achieved by integrating radical steps in more complex one-pot and domino processes.

References

1. F. W. Stacey and J. F. Harris, *Org. React.*, 1963, **13**, 150; C. Walling and M. S. Pearson, *Top. Phosphorus Chem.*, 1966, **3**, 1; W. G. Bentrude, *Acc. Chem. Res.*, 1982, **15**, 117; W. G. Bentrude, in *The Chemistry of Organophosphorus Compounds*, vol. 1, ed. F. R. Hartley, Wiley, Chichester, 1990, pp. 531–566.
2. D. Leca, L. Fensterbank, E. Lacote and M. Malacria, *Chem. Soc. Rev.*, 2005, **34**, 858–865.
3. C. Chatgilialoglu, V. I. Timokhin and M. Balestri, *J. Org. Chem.*, 1998, **63**, 1327.
4. S. Jockusch and N. J. Turro, *J. Am. Chem. Soc.*, 1998, **120**, 11773.
5. M. Anpo, R. Sutcliffe and K. U. Ingold, *J. Am. Chem. Soc.*, 1983, **105**, 3580.
6. M. Geoffroy and E. A. C. Lucken, *Mol. Phys.*, 1971, **22**, 257; C. M. L. Kerr, K. Webster and F. Wlliams, *J. Phys. Chem.*, 1975, **79**, 2650.
7. G. W. Sluggett, P. F. McGarry, I. V. Koptyug and N. J. Turro, *J. Am. Chem. Soc.*, 1996, **118**, 7367.
8. T. Sumiyoshi and W. Schnabel, *Macromol. Chem.*, 1985, **186**, 1811.
9. D. H. R. Barton, D. O. Jang and J. C. Jaszberenyi, *J. Org. Chem.*, 1993, **58**, 6838, and references cited therein; D. O. Jang, *Tetrahedron Lett.*, 1996, **37**, 5367.
10. D. O. Jang, D. H. Cho and D. H. R. Barton, *Synlett.*, 1998, **39**.
11. D. O. Jang, D. H. Cho and J. Kim, *Synth. Commun.*, 1998, **28**, 3559.

12. C. G. Martin, J. A. Murphy and C. R. Smith, *Tetrahedron Lett.*, 2000, **41**, 1833., R. McCague, R. G. Pritchard, R. J. Stoodley and D. S. Williamson, *Chem. Commun.*, 1998, 2691.
13. H. Nambu, G. Anilkumar, M. Matsugi and Y. Kita, *Tetrahedron*, 2002, **59**, 77.
14. D. H. Cho and D. O. Jang, *Tetrahedron Lett.*, 2005, **46**, 1799.
15. T. A. Khan, R. Tripoli, J. J. Crawford, C. G. Martin and J. A. Murphy, *Org. Lett.*, 2003, **5**, 2971, and references cited therein.
16. Y. Kita, H. Nambu, N. G. Ramesh, G. Anilkumar and M. Matsugi, *Org. Lett.*, 2001, **3**, 1157–1160.
17. V. T. Perchyonok, K. L. Tuck, S. J. Langford and M. W. Hearn, *Tetrahedron Lett.*, 2008, **49**(32), 4777–4779.
18. V. T. Perchyonok, S. Zhang and J. Chen, *Lett. Org. Chem.*, 2008, **5**, 304–309.
19. M. T. Reding and T. Fukuyama, *Org. Lett.*, 1999, **1**, 973–976.
20. A. E. Graham, A. V. Thomas and R. Yang, *J. Org. Chem.*, 2000, **65**, 2583.
21. H. Yorimitsu, H. Shinokubo and K. Oshima, *Bull. Chem. Soc. Jpn.*, 2001, **74**, 225–235.

CHAPTER 6

Organoboron Compounds as Radical Reducing Agents

The ability of organoboron compounds to participate in free radical reactions has been identified since the earliest investigation of their chemistry.[1] For instance, the autoxidation of organoboranes (Scheme 6.1) has been proved to involve radical intermediates.[2] This reaction led to the use of triethylborane as a universal radical initiator functioning under a very wide range of reaction conditions (temperature and solvent).[3]

Interestingly, homolytic substitution at boron does not proceed with carbon-centred radicals.[4] However, many different types of heteroatom-centred radicals, for example alkoxyl radicals, react efficiently with organoboranes (Scheme 6.2).

This difference in reactivity is caused by the Lewis base character of the heteroatom-centred radicals. Indeed, the first step of the homolytic substitution is the formation of a Lewis acid–Lewis base complex between the borane and the radical. This complex can then undergo a β-fragmentation leading to the alkyl radical. This process is of particular interest for the development of radical chain reactions.

The chapter will focus of the use of organoboron compounds in radical chemistry and concentrate on applications where the organoborane is used as an initiator, as a direct source of carbon-centred radicals, as a chain transfer reagent and finally as a radical reducing agent in aqueous media. The simple formation of carbon–heteroatom bonds *via* a radical process is not covered in this chapter as it has been treated in previous review articles.[5]

6.1 On the Use of Boron in Atom Transfer Processes

6.1.1 Iodine Atom Transfer

Triethylborane in combination with oxygen provides an efficient and useful system for iodine atom abstraction from alkyl iodides and therefore is a good initiator for iodine atom transfer reactions.[6] Indeed, the ethyl radical, issuing

RSC Green Chemistry No. 6
Radical Reactions in Aqueous Media
By V. Tamara Perchyonok
© V. Tamara Perchyonok 2010
Published by the Royal Society of Chemistry, www.rsc.org

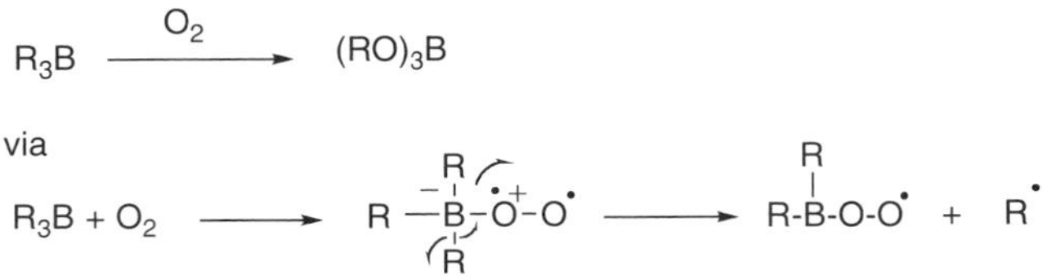

Scheme 6.1   Autoxidation of organoboranes.

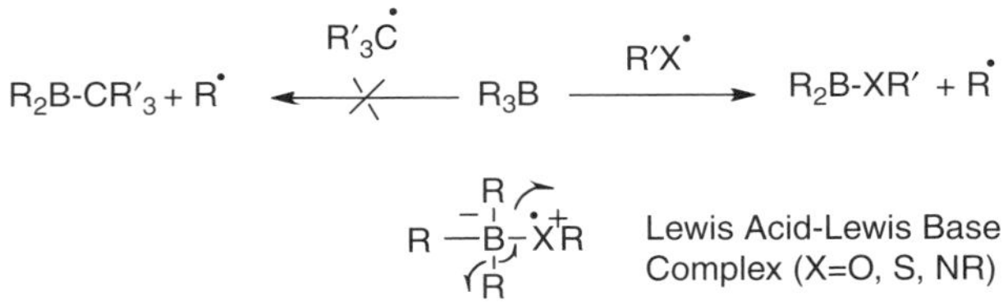

Scheme 6.2   Reactivity of carbon- and heteroatom-centred radicals towards organoboranes.

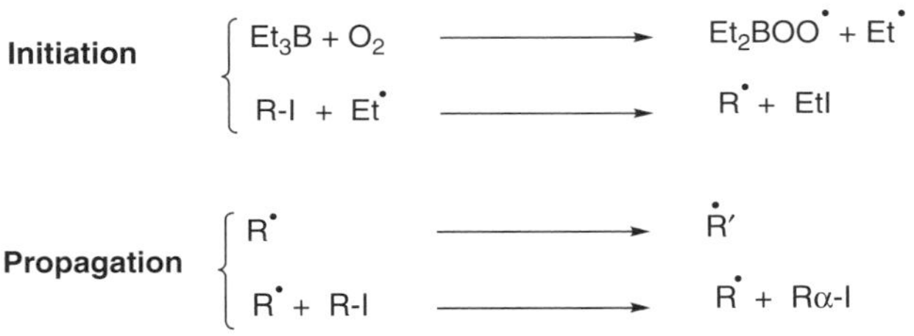

Scheme 6.3   Mechanism of Et$_3$B-mediated iodine transfer reaction.

from the reaction of triethylborane with molecular oxygen, can abstract an iodine atom from the radical precursor to produce a radical R$^{\bullet}$ that enters into the chain process (Scheme 6.3). The iodine exchange is fast and efficient when R$^{\bullet}$ is more stable than the ethyl radical.

Et$_3$B-induced additions of perfluoroalkyl iodides,[7] α-iodo esters (Scheme 6.4), iodoamides,[8] α-iodonitriles[9] and simple alkyl iodides to alkenes and alkynes have been reported. Interestingly, these reactions were also performed with success in aqueous media,[10] demonstrating the ability of Et$_3$B to act as an initiator in water.

Triethylborane is also an excellent initiator for intramolecular iodine atom transfer reactions. For example, cyclisation of the propargyl α-iodoacetal depicted in Scheme 6.5 gives the corresponding bicyclic vinyl iodide in high yield.[10] Allyliodoacetamides (Scheme 6.5) and allyl iodoacetates (Scheme 6.5) cyclise cleanly under Et$_3$B–O$_2$ initiation. In the case of the ester, the reaction has to be run in refluxing benzene in order to allow Z/E-ester isomerisation prior to

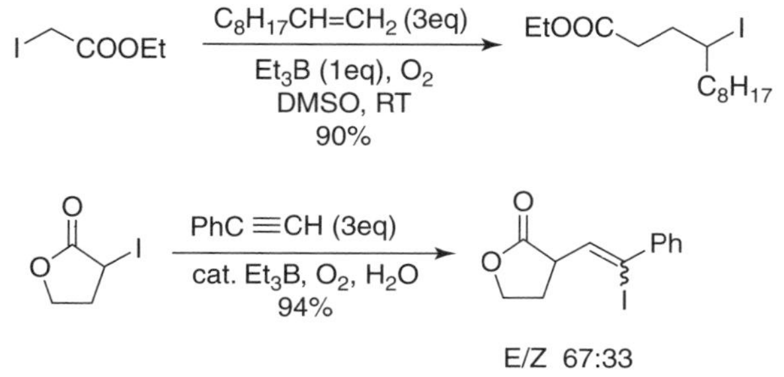

Scheme 6.4 Intermolecular addition reactions through iodine atom transfer.

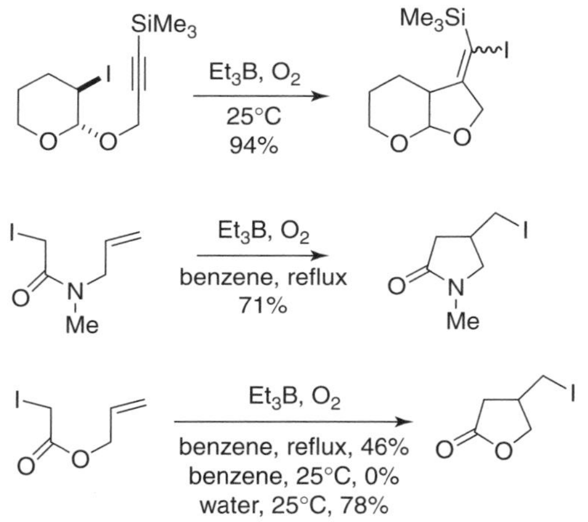

Scheme 6.5 Cyclisations through iodine atom transfer.

cyclisation.[11] No trace of cyclised product is detected when the reaction is carried out at room temperature. Interestingly, by running the same reaction in water, Oshima obtained the desired lactone in 78% yield. It was suggested that water facilitates the Z/E isomerisation. Efficient preparation of medium- and large-ring lactones in water has also been reported.[12,13]

6.1.2 Bromine Atom Transfer

Bromides are less reactive than the corresponding iodides in atom transfer processes. However, activated bromides such as diethyl bromomalonate and bromomalonitrile[13] react with alkenes under Et_3B-O_2 initiation. Kharasch-type reactions of bromotrichloromethane with alkenes are also initiated by Et_3B-O_2.[14] On the other hand, a remarkable Lewis acid effect was reported by Porter.[15] Atom transfer reactions of an α-bromooxazolidinone amide with

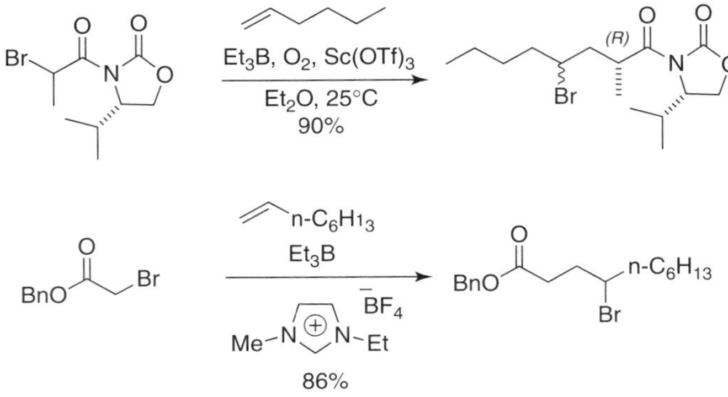

Scheme 6.6   Bromine atom transfer reactions in organic and alternative media.

alkenes are strongly favoured in the presence of Lewis acids such as $Sc(OTf)_3$ or $Yb(OTf)_3$, and this reaction was successfully applied to the diastereoselective alkylation of chiral oxazolidinone derivatives (Scheme 6.6). More recently, Oshima reported that bromine atom transfers take place at room temperature in ionic liquid media (Scheme 6.6).[16]

6.2   Organoboranes as Chain Transfer Reagents

In the preceding chapters it was noted that trialkylboranes are useful radical initiators as well as an efficient source of alkyl radicals. Organoboranes can also be used as chain transfer reagents. This approach is used when the direct reaction between the radical precursor and the radical trap cannot proceed (Scheme 6.7). Alkyl radicals generated from the organoboranes are not involved in product formation, but they produce the radicals leading to products. For this purpose, an extra step such as an iodine atom transfer or a

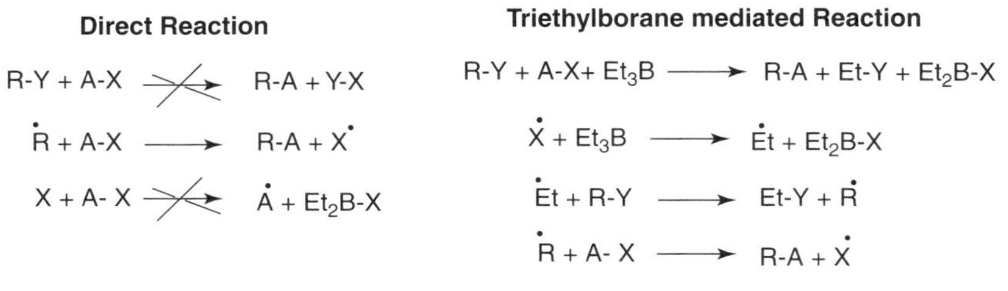

Where: R-Y = radical precursor, R-A = desired product, A-X = Radical Trap, X = Heteroatom centered radical, R = radical involved in product formation.

Scheme 6.7   Triethylborane as a chain transfer reagent for the conversion of R–Y to R–A.

hydrogen abstraction is necessary. This point is illustrated schematically in Scheme 6.7 for a triethylborane-mediated process. This reaction takes advantage of the high affinity of trialkylboranes for heteroatom-centred radicals X•.

6.2.1 Triethylborane-mediated Chain Transfer Reaction via Iodine Atom Transfer

Radical addition of organoboranes to imines and related compounds is a promising alternative to the use of classical organometallic compounds. However, this approach is limited to the few trialkylboranes that are readily available and cheap since only one of the three alkyl group is transferred. By using a triethylborane as a chain transfer reagent, the reaction could be extended to alkyl iodides as radical precursors. Bertrand[16] and Naito[17] both reported the use of triethylborane for the tin-free addition of alkyl iodides to imines.

The radical carboazidation of alkenes has been achieved in water using triethylborane as initiator.[18] This efficient process is complete in 1 h at room temperature in an open-to-air reaction vessel (Scheme 6.8). These new tin-free carboazidation conditions are environmentally friendly and allow reactions to be run with an excess of either the alkene or the radical precursor. They are also suitable for simple radical azidation of alkyl iodides and for more complex cascade reactions involving annulation processes. In both reactions, an excess of triethylborane (3 equiv) is required to obtain a good yield. This may be an indication that the chain process, or more precisely the reaction between the phenylsulfonyl radical and Et_3B, is not efficient.

6.3 Organoboron Compounds as Radical Reducing Agents as Complexes with Water and Alcohols

Wood[19] reported an innovative development of the Barton–McCombie deoxygenation of alcohols that allowed it to work under tin-free conditions (Scheme 6.9). A trimethylborane–water complex proved to be an efficient

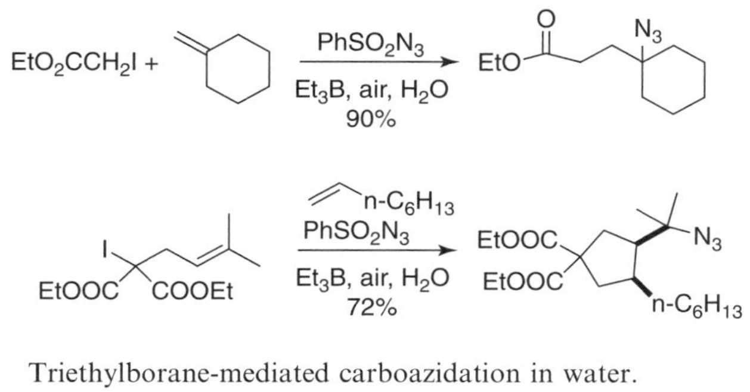

Scheme 6.8   Triethylborane-mediated carboazidation in water.

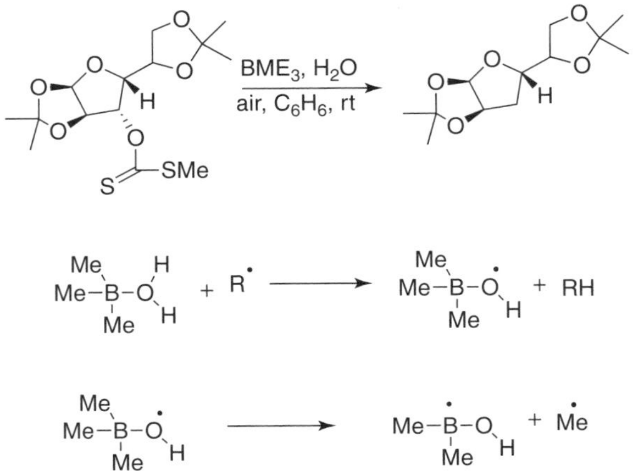
BME3, H2O
air, C6H6, rt
SMe
Me
Me–B–O
H
R
RH

Scheme 6.9 Barton–McCombie deoxygenation with Me_3B–water complex.

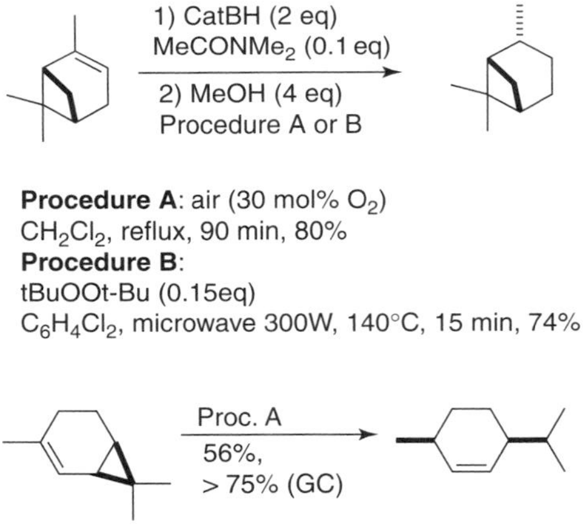
1) CatBH (2 eq)
MeCONMe2 (0.1 eq)
2) MeOH (4 eq)
Procedure A or B
Procedure A: air (30 mol% O2)
CH2Cl2, reflux, 90 min, 80%
Procedure B:
tBuOOt-Bu (0.15eq)
C6H4Cl2, microwave 300W, 140°C, 15 min, 74%
Proc. A
56%,
> 75% (GC)

Scheme 6.10 Mild radical-mediated reduction of organoboranes with methanol.

reagent for the reduction of xanthates. Complexation of water by trimethylborane induces a strong decrease in the O–H bond dissociation energy from 116 kcal mol^{-1} (water) to 86 kcal mol^{-1} (Me_3B–water complex).

Renaud made a similar observation and reported a mild and efficient radical-mediated reduction of organoboranes (Scheme 6.10).[20] An *in situ*-generated *B*-methoxycatecholborane–methanol complex acts as a reducing agent. The radical nature of the process was demonstrated by using (+)-2-carene as a radical probe. Water, ethanol and trifluoroethanol can be used instead of MeOH with very similar efficiency.

The reaction mechanism of this transformation involves activation of the O–H bond of methanol by complexation with *B*-methoxycatecholborane.

Interestingly, the reduction leads after fragmentation of the radical–ate complex to a methoxyl radical that reacts very efficiently with the *B*-alkyl-catecholborane, leading to an efficient chain process.

6.4 Conclusion and Future Directions

The tremendous development of the use of radicals in organic synthesis and the necessity of avoiding the use of tin derivatives because of their toxicity have led to a revival of the radical chemistry of organoboranes in water and aqueous media. The use of triethylborane as an initiator for radical chain reactions is now part of the classical arsenal of organic chemists. The generation of more complex and functionalised radicals from organoboranes is of great interest since it allows one to consider alkenes as a potential source of radicals. So far, the generation of radicals has not been extended to alkenyl and aryl radicals, but rapid progress is expected in this field. Interestingly, organoboranes could also play the role of chain transfer reagents in radical processes. Due to the particularly high reactivity of boron derivatives, the design of tandem processes involving radical and non-radical reactions is now possible. Finally, boron derivatives are promising reagents for activating water and alcohols and making them suitable reagents for the reduction of radicals. Spectacular development in this particular field is expected in the near future in aqueous and alternative media, such as ionic liquids, microwaves and the solid state.

References

1. H. C. Brown and M. M. Midland, *Angew. Chem. Int. Ed. Engl.*, 1972, **11**, 692; A. Ghosez, B. Giese, T. Gšbel and H. Zipse, in *Methods of Organic Chemistry*, 4th Ed., Houben-Weyl, Stuttgart, 1995, Vol. E21d, p. 3913–3943. C. Ollivier and P. Renaud, *Chem. Rev.*, 2001, **101**, 3415.
2. A. G. Davies and B. P. Roberts, *Acc. Chem. Res.*, 1972, 5, 387; A. G. Davies, *Pure Appl. Chem.*, 1974, **39**, 497.
3. K. Nozaki, K. Oshima and K. Utimoto, *J. Am. Chem. Soc.*, 1987, **109**, 2547; H. Yorimitsu and K. Oshima, in *Radicals in Organic Synthesis*, eds. P. Renaud and M. P. Sibi, Wiley-VCH, Weinheim, 2001 **Vol. 1**.
4. R. A. Batey and D. V. Smil, *Angew. Chem. Int. Ed.*, 1999, **38**, 1798.
5. A.-P. Schaffner and P. Renaud, *Eur. J. Org. Chem.*, 2004, 2291.
6. H. Yorimitsu, H. Shinokubo and K. Oshima, *Synlett.*, 2004, 674; D. P. Curran, M.-H. Chen, E. Spetzler, C. M. Seong and C.-T. Chang, *J. Am. Chem. Soc.*, 1989, **111**, 8872.
7. E. Baciocchi and E. Muraglia, *Tetrahedron Lett.*, 1994, **35**, 2763.
8. Y. Tang and C. Li, *Org. Lett.*, 2004, **6**, 3229.
9. Y. Takeyama, Y. Ichinose, K. Oshima and K. Utimoto, *Tetrahedron Lett.*, 1989, **30**, 3159.
10. Y. Ichinose, S.-I. Matsunaga, K. Fugami, K. Oshima and K. Utimoto, *Tetrahedron Lett.*, 1989, **30**, 3155.

11. M. Ikeda, H. Teranishi, N. Iwamura and H. Ishibashi, *Heterocycles*, 1997, **45**, 863; M. Ikeda, H. Teranishi, K. Nozaki and H. Ishibashi, *J. Chem. Soc., Perkin Trans. 1*, 1998, 1691.

12. Y. Kita, A. Sano, T. Yamaguchi, M. Oka, K. Gotanda and M. Matsugi, *Tetrahedron Lett.*, 1997, **38**, 3549.

13. H. Yorimitsu and K. Oshima, *Bull. Chem. Soc. Jpn.*, 2002, 75.

14. T. Nakamura, H. Yorimitsu, H. Shinokubo and K. Oshima, *Synlett.*, 1998, 1351.

15. C. L. Mero and N. A. Porter, *J. Am. Chem. Soc.*, 1999, **121**, 5155.

16. M. Bertrand, L. Feray and S. Gastaldi, *C. R. Acad. Sci. Chim.*, 2002, **5**, 623; M. P. Bertrand, L. Feray, R. Nouguier and L. Stella, *Synlett.*, 1998, 780.

17. H. Miyabe M. Ueda and T. Naito, *Synlett.*, 2004, 7, 1140; H. Miyabe, R. Shibata, M. Sangawa, C. Ushiro and T. Naito, *Tetrahedron*, 1998, **54**, 11431.

18. P. Panchaud and P. Renaud, *J. Org. Chem.*, 2004, **69**, 3205.

19. D. A. Spiegel, K. B. Wiberg, L. N. Schacherer, M. R. Medeiros and J. L.Wood, *J. Am. Chem. Soc.*, 2005, **127**, 12513–12515; M. R. Medeiros, L. N. Schacherer, D. A. Spiegel and J. L. Wood, *Org. Lett.*, 2007, **9**, 4427–4429; J. Jin and M. Newcomb, *J. Org. Chem.*, 2007, **72**, 5098–5103.

20. D. Pozzi, E. M. Scanlan and P. Renaud, *J. Am. Chem. Soc.*, 2005, **127**, 14204.

CHAPTER 7

Carbon–Carbon Bond Formation through Radical Addition Chemistry

7.1 General Principles and Considerations

The most important methodology for the aliphatic C–C bond formation *via* radical reactions is the addition of the radical to an alkene double bond, both inter- and intramolecularly (with the 5-*exo*-ring cyclisation mode preferred in the latter case). This reaction leads to adduct radicals that must be converted to non-radical products before polymerisations can take place. For this reason, polymerisation is avoided either by intermolecular trapping of adduct radicals or by intramolecular, homolytic bond cleavage. Hydrogen atom donors X–H, heteroatom donors X–Z or electron donors M^{n+} are used as trapping agents (Scheme 7.1).

Relatively straightforward but extremely useful hydrogen atom transfer (HAT) reactions involve the replacement of various X groups with hydrogen without alterations to the carbon skeleton. If the intermediate carbon radical could be intercepted by a C–C bond forming process before the HAT reaction

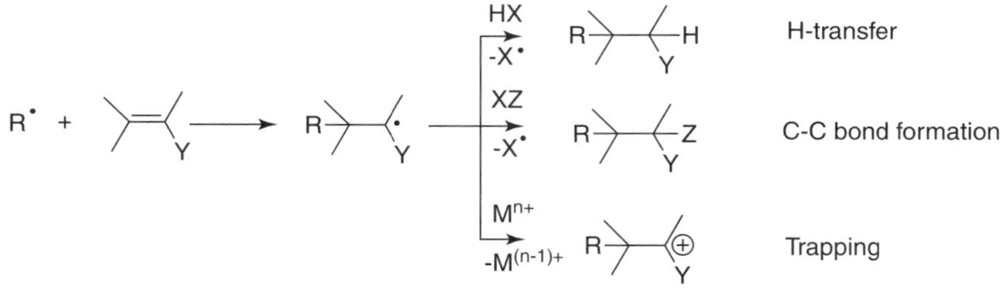

Scheme 7.1 General mechanistic considerations for radical additions to C=C bonds.

RSC Green Chemistry No. 6
Radical Reactions in Aqueous Media
By V. Tamara Perchyonok
© V. Tamara Perchyonok 2010
Published by the Royal Society of Chemistry, www.rsc.org

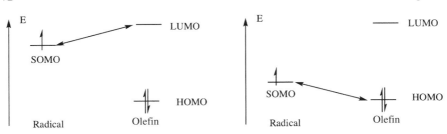

Radical addition to electron-poor alkenes   Radical addition to electron-rich alkenes

Scheme 7.2   Radical additions to electron-poor and electron-rich olefins.

takes place, then the architecture of the substrate could be profoundly modified and its complexity rapidly increased. The factors that play an important role in the outcomes of the radical additions to alkenes can be broadly classed as polar and steric effects (Scheme 7.2).[1] Often the interaction between two molecules can be expressed as a combination of attractive and repulsive terms reflecting polar (electrostatic and orbital interactions) and steric effects (usually repulsive). However, in case of radicals, which are not charged species, the relative difference in the electrostatic (or coulombic) term is small or negligible and polar effects become dominated by orbital interactions. Frontier orbitals thus play a prominent role and the reactivity can indeed often be understood, at a qualitative level, by simply looking at the energy gap between the SOMO of the radical and the LUMO or the HOMO of the olefin (Scheme 7.2).[2]

7.2 'Metal-free' Radical Addition to C=C Double Bonds in Water or Aqueous Media

The addition of carbon-based radicals has been shown to be successful in water.[3] Thus, radical additions of a 2-iodoalkanamide or 2-iodoalkanoic acid to alkenols using the water-soluble radical initiator 4,4′-azobis(4-cyanopenta-noic acid) or 2,2′-azobis(2-methylpropanamidine) dihydrochloride carried out in water generated γ-lactones (Scheme 7.3).

The addition of perfluoroalkyl iodides to simple alkenes has been successful under aqueous conditions for the synthesis of fluorinated hydrocarbons.

In addition to carbon-based radicals, other radicals such as sulfur-based radicals, generated from an RSH-type precursor (R = alkyl, acyl) with AIBN also smoothly added to allylglycine.[4] Optimal results were obtained when both the unsaturated amino acids and RSH dissolved completely in the reaction medium (dioxane–water or methanol–water were found to be superior solvents). Radical additions of thiophenol to carbon–carbon multiple bonds and radical cyclisation of *N*-allyl-2-iodoalkanamide in aqueous media proceeded smoothly to the *N*-acetylpyrrolidine derivative in 96% yield. Under similar

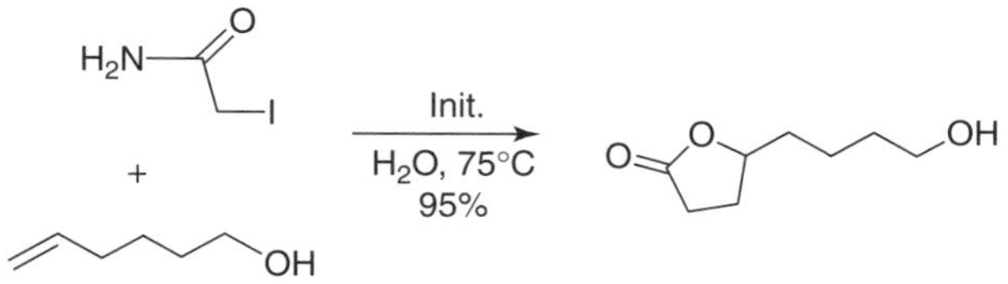

Scheme 7.3 Addition of 2-iodoacetarnide to an alkenol affording a γ-lactone.

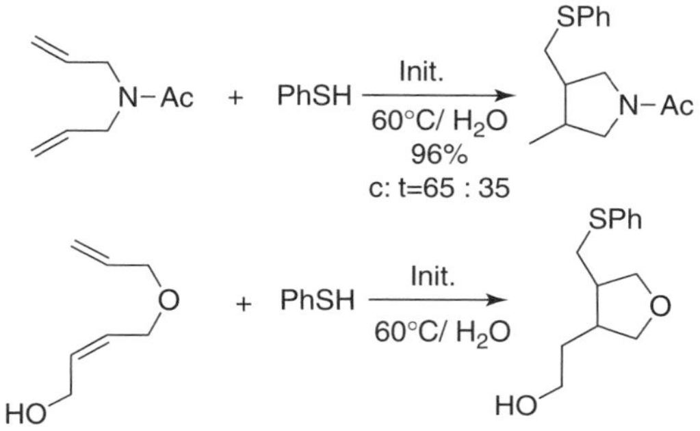

Scheme 7.4 Atom transfer radical cyclisation of *N,N*-diallylacetamide and (*Z*)-4-(allyloxy)-but-2-en-1-ol.

experimental conditions, atom transfer radical cyclisation of 2-iodoamides in water gave corresponding lactams in 96% and 85% yield (Scheme 7.4).[4]

Although a media effect in radical reactions was believed to be almost negligible, Oshima and co-workers reported that water increases the rate of radical reactions.[5,6] Water as a reaction solvent also markedly promotes the cyclisation of large-membered rings. Stirring a solution of 3,6-dioxa-8-nonenyl iodoacetate in water in the presence of triethylborane at 25 °C for 10 h provided the 12-membered ring product 4-iodo-6,9-dioxa-11-undecanolide in 84% yield, whereas the cyclisation in benzene afforded the lactone in only 23% yield (Scheme 7.5).[7] *Ab initio* calculations on the cyclisation indicated that the large dielectric constant of water lowers the barrier not only of the rotation from the *Z*-rotamer to the *E*-rotamer that can cyclise, but also of the cyclisation constructing the γ-lactone framework. Moreover, the high cohesive energy of water also causes acceleration of the cyclisation because water forces a decrease in the volume of the reactant (Scheme 7.5).[7] This observation is suggestive of the fact that water is not an inert solvent as previously thought.

In parallel, the oxime ethers are well known to be excellent radical acceptors.[8]

Naito and co-workers investigated aqueous medium radical addition to glyoxylic oxime ether due to its good reactivity in organic solvents.[9] Its noteworthy that triethylborane acts as a radical initiator and terminator to trap the intermediate alkyl radical (Scheme 7.6).

Recently the same group reported new free radical-mediated tandem reactions of oxime ethers for the synthesis of heterocycles *via* two C–C bond-forming

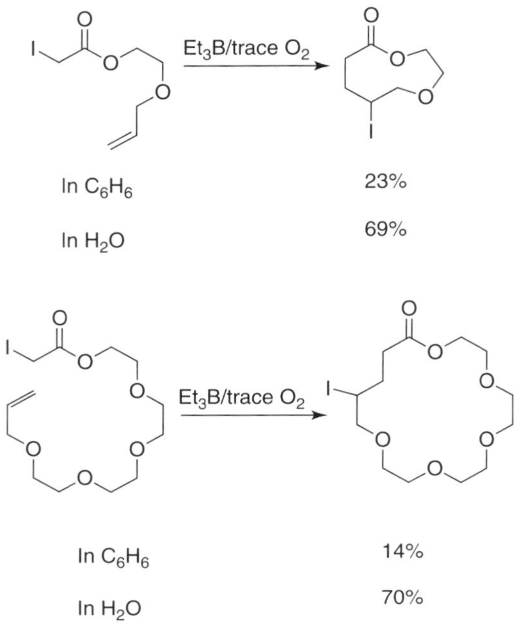

In C$_6$H$_6$

In H$_2$O

23%

69%

In C$_6$H$_6$

In H$_2$O

14%

70%

Scheme 7.5 Medium- and large-ring formation by radical cyclisation in water.

processes. In addition to radical cyclisation of oxime ethers, the tandem radical reactions revealed a broader aspect of the utility of oxime ethers as a radical acceptor for the synthesis of various types of amino compounds, as shown in Scheme 7.7.[10]

7.3   Indium-mediated Radical Reactions in Aqueous Medium

The utility of indium as a free radical initiator in aqueous media can be directly linked with the first ionisation potential (5.8 eV) and is as low as that of lithium and sodium. Therefore, it is well accepted that indium has the potential to induce radical reactions as a radical initiator *via* a single electron transfer process (Scheme 7.8).[11]

In 1991, Li and Chan reported the use of indium to mediate Barbier-Grignard-type reactions in water.[12] The work was designed on the basis of the first ionisation potentials of different elements, in which indium has the lowest first ionisation potential relative to the other metallic elements near it in the periodic table. On the other hand, indium metal is not sensitive to boiling water or alkali and does not form oxides readily in air. Such special properties of indium indicate that it is perhaps a promising metal for aqueous Barbier–Grignard-type reactions. Indeed, it appears that indium is the most reactive and

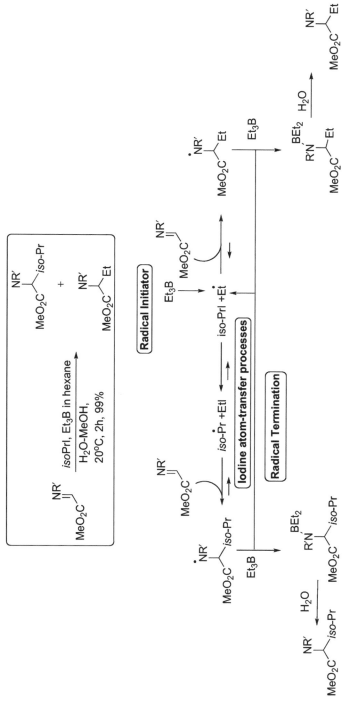

Scheme 7.6 Alkyl radical addition to glyoxylic oxime initiated by Et$_3$B and proposed mechanism.

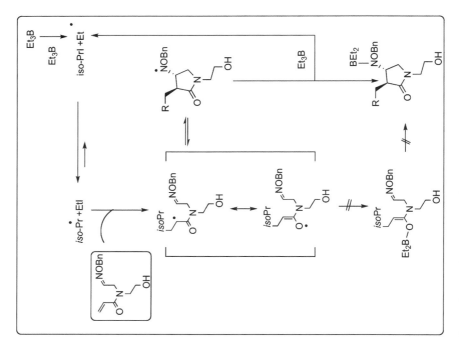

Scheme 7.7 Radical cyclisation of oxime ethers as radical precursors for various amino compounds.

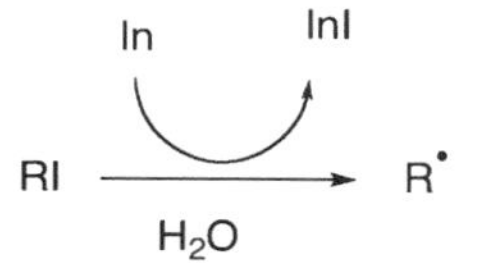

Scheme 7.8 Indium as a radical initiator in H_2O.

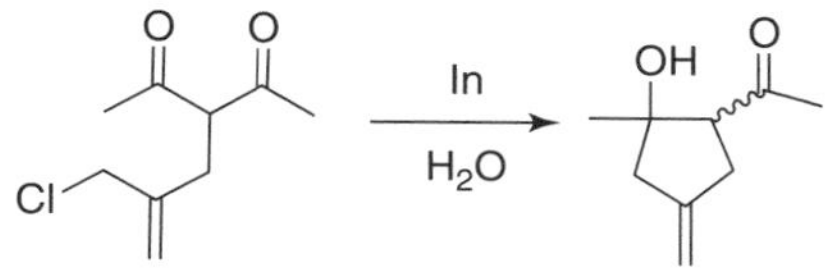

Scheme 7.9 Alkylation of 1,3-dicarbonyl compounds using allyl bromide or allyl chloride and indium in water.

effective metal for such reactions. When the allylation was mediated by indium in water, the reaction proceeded smoothly at room temperature without any promoter, whereas the use of zinc and tin usually requires acid catalysis, heat or sonication. A variety of 1,3-dicarbonyl compounds have been alkylated successfully using allyl bromide or allyl chloride in conjunction with either tin or indium (Scheme 7.9). The reaction can be used readily for the synthesis cyclopentane derivatives.[13]

Naito and co-workers studied the addition–cyclisation–trapping reaction in aqueous media mediated by indium.[14] Tandem C–C bond-forming reactions were studied using indium as a single electron transfer radical initiator. The radical addition–cyclisation–trapping reaction of a substrate having a vinyl-sulfonamide group and an alkene moiety preceded smoothly in aqueous media. The radical addition–cyclisation reaction of hydrazone gave the functionalised cyclic products, as shown in Scheme 7.10.

Intermolecular alkyl radical addition to imine derivatives was studied in aqueous media using indium as a single electron transfer radical initiator. The one-pot reaction based on radical addition to glyoxylic hydrazone provided a convenient method for preparing the α-amino acids. Indium-mediated radical addition to an electron-deficient C=C bond also proceeded effectively to provide a new carbon–carbon bond-forming method in aqueous media (Scheme 7.11).[15]

The carbon–nitrogen double bond has emerged as a radical acceptor and thus several intermolecular radical addition in organic solvents reactions have been investigated.[16] However, investigations by Naito and co-workers have shown that imine derivatives such as oxime ethers, hydrazones and nitrones are excellent water-resistant radical acceptors for aqueous medium reactions using triethylborane as radical initiator. The one-pot reaction based on radical addition to glyoxylic hydrazone was examined under several reaction conditions (Table 7.1).[17] The monophasic reaction of an imine in H_2O–MeOH gave

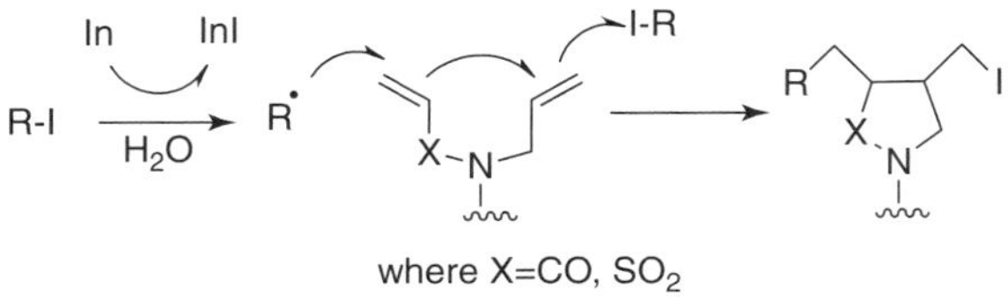

where X=CO, SO$_2$

Scheme 7.10 Indium-mediated tandem radical addition–cyclisation–trapping reactions in aqueous media.

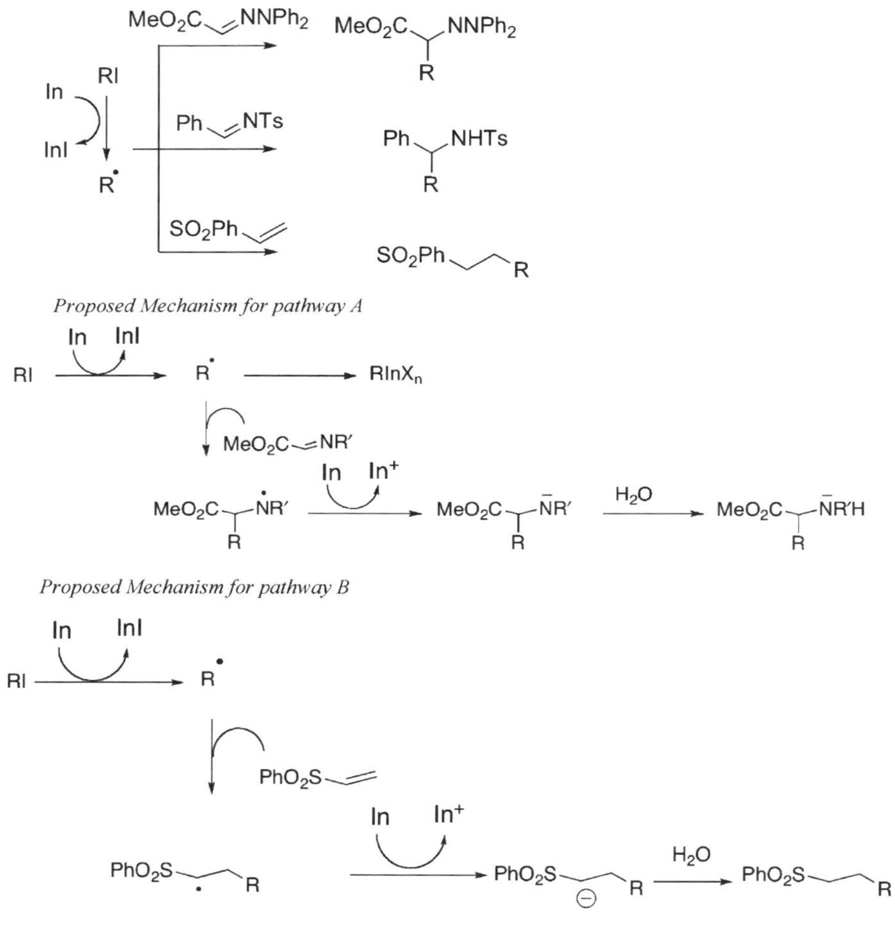

Proposed Mechanism for pathway A

Proposed Mechanism for pathway B

Scheme 7.11 Proposed mechanisms for indium-mediated C–C bond formation.

the isopropylated product after stirring for only 30 min, without formation of significant by-products such as reduced product.[17] It is also important to note that practically no reaction occurred in the absence of water. This observation suggests that water would be important for the activation of indium and for

Table 7.1 Alkylation reactions of various amines.

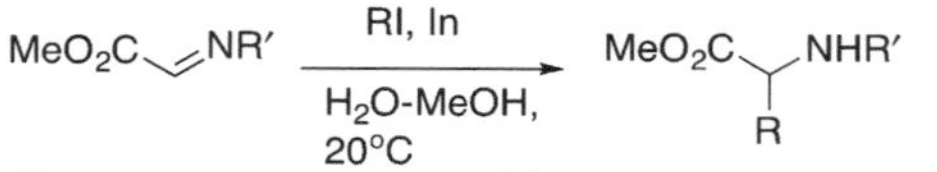

Entry	R'	R	Yield (%)
1	OBn	*i*-Pr	74
2	NPh$_2$	*i*-Pr	98
3	NPh$_2$	*c*-Pentyl	90

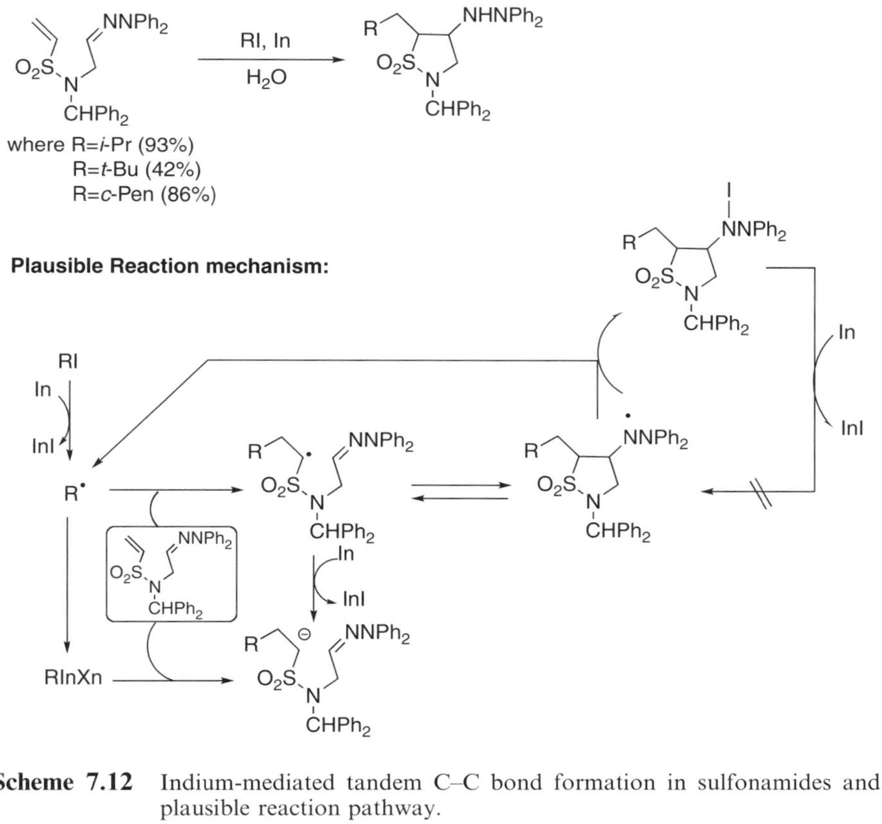

where R=*i*-Pr (93%)
R=*t*-Bu (42%)
R=*c*-Pen (86%)

Plausible Reaction mechanism:

Scheme 7.12 Indium-mediated tandem C–C bond formation in sulfonamides and plausible reaction pathway.

proton donation to the resulting amide anion. In the presence of a galvinoxyl free radical as radical scavenger, the reaction did not proceed effectively, suggesting that this reaction proceeds *via* a free radical mechanism.[18]

Another strategy involving tandem radical reactions that offers an advantage of multiple carbon–carbon bond formation in a single operation in the aqueous medium tandem construction of carbon–carbon bonds has been widely being studied by Naito and co-workers (Scheme 7.12). The indium-mediated tandem

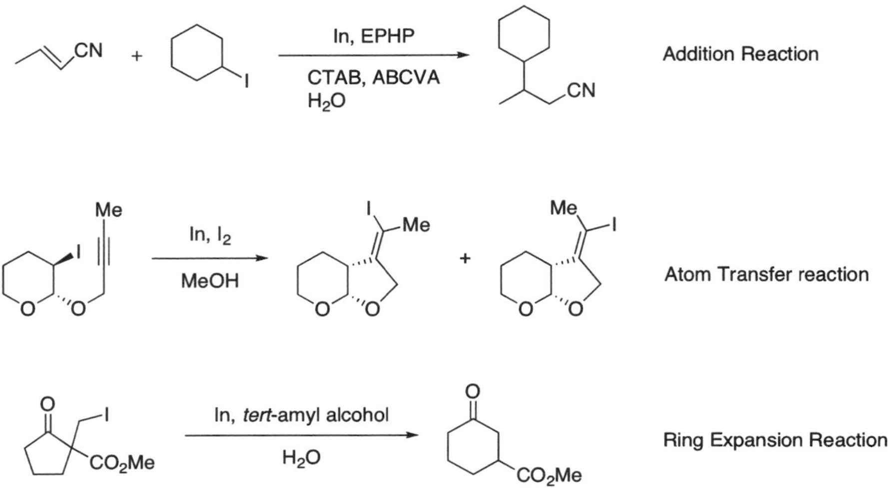

Scheme 7.13 Various radical reactions mediated by indium.

carbon–carbon bond-forming reaction of substrates having two different radical acceptors is shown in Scheme 7.12.[18] The tandem radical addition–cyclisation–trapping reactions in the presence of RI in water gave the cyclic products in good yields without the formation of other by-products.

More recently, the related indium-mediated radical reactions have been widely studied (Scheme 7.13).[19] Indium iodide-mediated radical cyclisation was first reported by Cook *et al.*[20] The indium-mediated 1,4-addition of alkyl radicals to (*E*)-but-2-enenitrile was investigated by using 1-ethylpiperidinum hypophosphite (EPHP) as a hydrogen donor in aqueous media (Scheme 7.13).[21] Atom transfer radical cyclisation and reductive radical cyclisation were studied using indium and iodine.[22] Indium mediated alkyl radical addition to dehydroamino acid derivatives was also reported.[23] The indium-mediated radical ring expansion of α-halomethyl cyclic β-keto esters, shown in Scheme 7.13, was achieved in aqueous alcohols.[24]

7.4 Zinc-mediated Alkylations of C=N Bonds: Yet Another Tool in the Carbon–Carbon Bond Formation Toolbox

Zinc-mediated carbon–carbon bond-forming reactions have attracted considerable interest in recent years.[12] Among the various types of known zinc-mediated reactions, radical reactions have captured much recent attention because of their exceptional tolerance to functional groups (Scheme 7.14).[25]

The carbon–nitrogen double bond of imine derivatives has been shown to act as a radical acceptor. For example, it was shown that *N*-sulfonylimines are

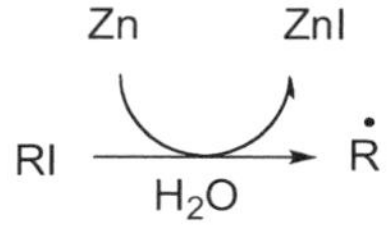

Scheme 7.14 Zinc as a radical initiator in H_2O.

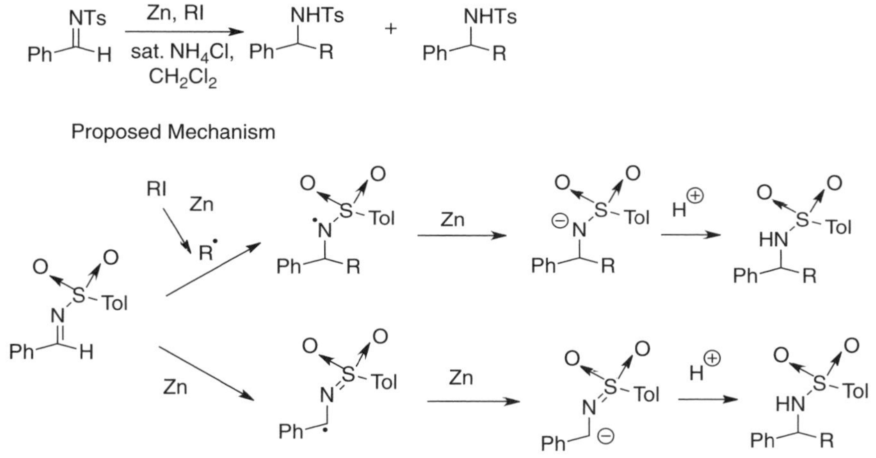

Scheme 7.15 Zinc-mediated alkyl radical addition to *N*-sulfonylimine.

excellent radical acceptors for the aqueous medium reaction using zinc as a single electron transfer radical initiator (Scheme 7.15).[25]

As part of environmentally benign synthetic reactions in aqueous media employing imines as a substrate, zinc-mediated radical additions to glyoxylic oxime ethers and hydrazones for asymmetric synthesis of α-amino acids have been reported. The zinc-mediated addition of isopropyl radical to achiral glyoxylic oxime ether (shown in Scheme 7.16) in aqueous medium (saturated NH_4Cl/aqueous MeOH) was examined. In that reaction isopropylated product was formed in 96% yield. It was suggested that this reaction proceeds through a radical pathway based on a single electron transfer from zinc.[26,27]

7.5 Stereoselective Radical Addition to C=N Bonds in Aqueous Media: How Does It Fit?

With a few notable exceptions, the majority of asymmetric reactions are performed in apolar and aprotic media, which precludes the use of water-soluble compounds. It is imperative that water, also, is fully explored as a reaction medium for asymmetric synthesis. Provided that substrates and reagents can be used that do not react with water, how will the selectivity of reactions in inert

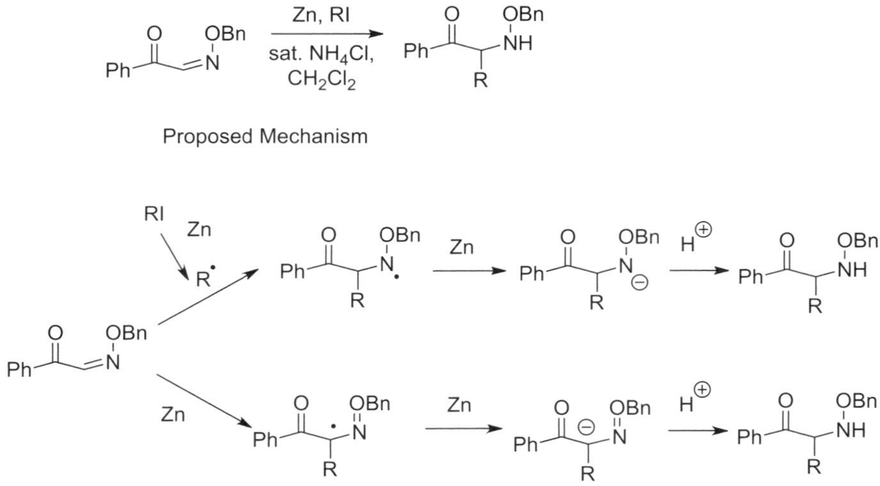

Proposed Mechanism

Scheme 7.16   Zinc-mediated addition to the C=N bond in aqueous media.

solvents be affected by the progression to a participating solvent such as water? Learning how to take advantage of the uniquely complex solvating properties of water may lead to new concepts and possibilities in asymmetric synthesis. Initial work in this area has indeed led to some interesting and surprising results.

In general, diastereoselectivity and enantioselectivity in free radical transformations can be achieved by using the following modes of stereoselectivity controls:[28-30]

1. A chiral Lewis acid can be used to bind to substrate or radical species and determine the approach of the other reacting component while accelerating the chiral pathway relative to the background reaction.
2. A chiral catalyst can be coordinated temporarily to the substrate and/or the reacting radical and bring about the reaction in an intramolecular sense.
3. A chiral chain-transfer agent can be used: this will determine the approach in an atom-transfer step.
4. A chiral environment can be provided, as in the photolysis of chiral crystals.
5. A chiral inductor along with the substrate can be trapped in an organised medium.
6. The chirality inherent in the molecule can be converted to molecular chirality. (Memory of chirality.)

Although reasonably well utilised in stereospecific free radical chemistry in conventional organic solvents the following principles are in their infancy in being applied and utilised in aqueous media. The examples discussed in detail will focus on the two most common means of stereocontrol, reactant control and reagent control, with diastereo- and enantioselective examples in free radical addition and hydrogen transfer reactions in aqueous media.

Let us rehearse some principles of stereoselective radical reactions in general. Stereoselective synthesis concerns the creation of a new stereogenic (chiral) centre by conversion of a prochiral substrate (*i.e.* radical) into a chiral product.[29] This transformation to be achieved as result of utilising:

- an appropriate chiral reagent;
- asymmetric induction by a stereocentre already present in the substrate (chiral auxiliary);
- a chiral complexing agent (*e.g.* a Lewis acid).

These reactions are diastereoselective if either the reagent or substrate contains a stereogenic centre. They are enantioselective if transfer of chirality occurs in the bond-forming step.

The major cause of difficulty in achieving stereoselectivity in free radical reactions lies within the nature of the prochiral carbon-centred radical. Experimental evidence suggests that prochiral radicals are almost planar and essentially sp^2 hybridised, with the unpaired electron occupying an unhybridised p-orbital located perpendicular to the plane of substitution (Figure 7.1).[30]

7.6 Diastereoselective Radical Reactions in Acyclic Systems

In general, diastereoselective reactions are those in which the radical undergoes a selective reaction with a prochiral radical trap. Control of stereoselective outcome in these reactions to be achieved through:

- substrate control
- auxiliary control
- complex control (in the presence of a Lewis acid).
- hydrogen bonding

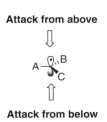

Attack from above

Attack from below

Figure 7.1 A prochiral radical being attacked from above or below during a stereoselective radical transformation.

7.7 Diasteroselective Free Radical Reactions Under Substrate-controlled Conditions

For any type of asymmetric inductions in acyclic systems to be successful, the following conditions need to be met:

1. The number of energetically favourable conformations of the diastereotopic transition state needs to be restricted such that a preference of one conformation over others is achieved.
2. The conformation should be such that the different substituents on the controlling stereocentre differentiate between the diastereofaces of the reacting prochiral radical.

Although many tools have been developed for use in the prediction of the outcomes of ionic reactions, some are also applicable to the prediction of the outcomes of free radical reactions. One of the most widely used ionic reactions in diastereoselective synthesis is addition of a nucleophile (Nu^-) to the carbonyl group of chiral compound R–CO–CLMS, where L, M and S represent, respectively, the largest, the middle-sized and the smallest of the substituents (other than the keto group) on the asymmetric carbon (Scheme 7.17).

Although Cram's rule is useful, its capability is not extensive.[31] Felkin, Ahn and co-workers subsequently developed the following model which accounts more fully for the chiral induction observed in many stereoselective transformations.[32] The difference between the Felkin–Ahn (FA) model and Cram's rule is in the conformation of the carbonyl group. In the FA model, group L is assumed not to be antiperiplanar to the carbonyl group, but positioned at a right-angle (Figure 7.2). The stereoselectivity of the transformation arises from the relative energies of the two transition states, which in turn depend on the interaction between M and R. The stereochemistry of the major product (*i.e.* alcohol) will result from the approach of the nucleophile from the side remote from L (path a). The minor product of the reaction is obtained through path b.

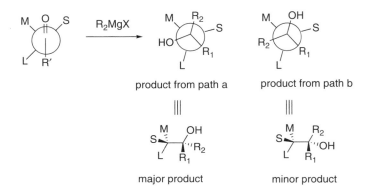

Scheme 7.17   Application of Cram's rule to a typical Grignard reaction.

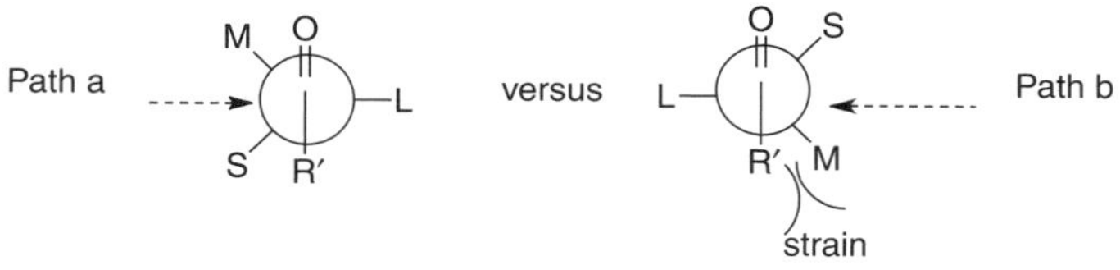

Figure 7.2 Felkin–Ahn model in action.

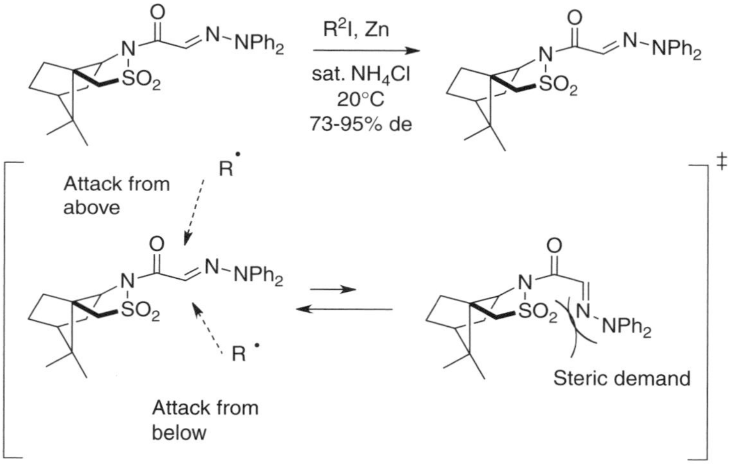

Scheme 7.18 Stereoselective radical addition to the C=N bond mediated by Zn.

As the carbon atom is sp^2 hybridised, the carbonyl group is planar with a carboxy oxygen and the two substituents oriented in a trigonal planar geometry which is almost identical with the configuration of a prochiral radical. Based on structural similarities between the carbonyl group and a prochiral radical, the predictive tools useful in ionic reactions can also be applied in radical transformations.[33–35] As a rule of thumb: Cram's rule is useful for substrates containing COR, CO$_2$R, CONR, AR, NR$_2$ and NO$_2$ substituents at the radical centre. Reactions of radicals with heteroatoms that are directly attached to the radical centre (OR, SR, NR) follow the Felkin–Ahn model.

7.8 Stereoselectivity in Free Radical Alkylation of C=N Bonds in Detail

Diastereoselective radical additions to hydrazones have been examined under several reaction conditions. The biphasic reaction of a hydrazone with an isopropyl radical in aqueous NH$_4$Cl–CH$_2$Cl$_2$ (4:1, v/v) proceeded slowly to give the corresponding alkyl-substituted derivative in 73% yield and 95% de after being stirred for 22 h (Scheme 7.18).[27] As expected, the diastereoselectivity

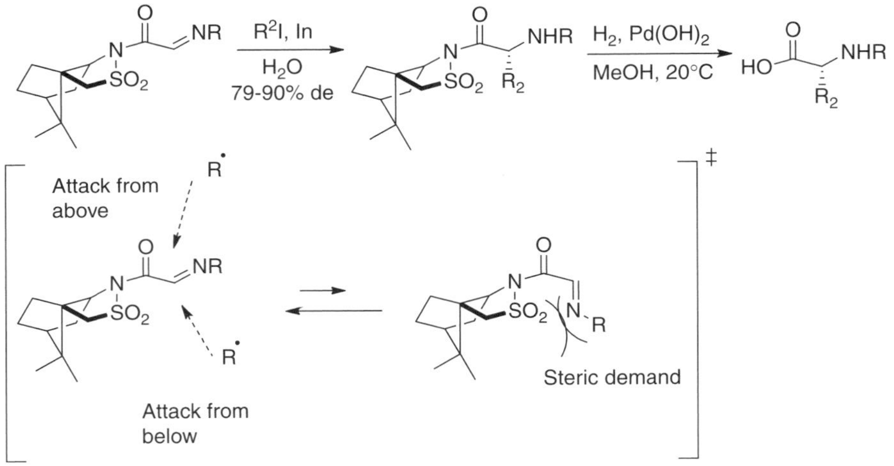

Scheme 7.19   Stereoselective radical addition to the $C=N$ bond mediated by indium towards the synthesis of amino acids.

observed can be explained as being due to the rotamer having the carbonyl group *anti* to the sulfonyl group being favoured in order to minimise dipole–dipole interactions between these groups (Scheme 7.17), as suggested by studies on the camphosultam derivatives of glyoxylic acid.[26,27]

Furthermore, the isopropylated product has been converted into α-amino acid (Scheme 7.19) through cleavage of the N–N bond of the diastereochemically pure product by hydrogenolysis in the presence of Pearlman's catalyst. In earlier work on triethylborane-induced radical reactions in organic solvents, oxime ether showed excellent reactivity. Thus we expect that the direct comparison of indium-mediated reactions with triethylborane-induced reactions would lead to informative and instructive suggestions regarding the reactivity and stereochemical course of the transformation in question.

7.9   Enantioselective Radical Addition Reactions to the C=N Bond Utilising Chiral Quaternary Ammonium Salts of Hypophosphorous Acids in Aqueous Media

Jang and co-workers reported on the development of an enantioselective radical addition reaction to glyoxylate oxime ether for the preparation of α-amino acids under mild reaction conditions with chiral quaternary ammonium salts of hypophosphorous acid in aqueous media.[26] The newly prepared chiral quaternary ammonium hypophosphites are inexpensive, less toxic than metal-containing compounds and the reaction conditions and workup are mild and simple (Table 7.2). It is also important to note that chiral quaternary hypophosphites are recyclable without altering their performance. The

Table 7.2 Radical addition reactions to glyoxylic oxime ether under various reaction conditions.

Entry	RI	HD	Yield[a] (%)	Selectivity[b] R:S
1	iPr-I	QP	83 (7)	21:79
2	iPr-I	QDP	82 (10)	62:38
3	1-Ad-I	QP	45 (35)	>1:99
4	1-Ad-I	QDP	47 (37)	>99:1
5	"Oct-I	QP	50 (25)	40:60
6	"Oct-I	QDP	48 (30)	58:42

[a]The yields in parentheses are for side product **2b**.
[b]The enantiomeric ratio was determined by HPLC analysis using a chiral column (Daicel Chiralpack AD-H) with hexane–2-propanol as the solvent.

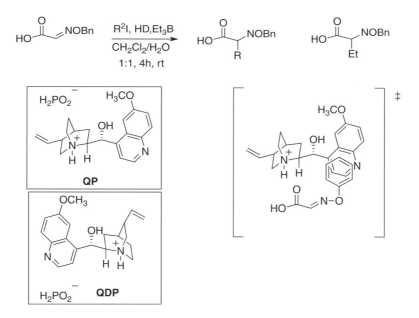

Scheme 7.20 Enantioselective addition reactions to glyoxylic oxime ether.

enantioselectivities afforded high yields of the addition products with a high enantioselectivity that can be attained without using metals (Scheme 7.20). The absolute configuration of the addition adducts suggested a plausible rationalisation for the observed *si*-face attack of the alkyl radical on the substrate involving p-stacking and hydrogen bonding. The substrate is bound to QP by hydrogen bonds between N–H and C=O/C=N with a CO/CN *s-cis* planar conformation and also through p-stacking. In this arrangement, the *re*-face of the C=N bond is blocked by the quinoline ring of QP. An alkyl radical then attacks the C=N bond from the *si*-face to afford the addition products.

7.10 Conclusion and Future Directions

Recent advances in free radical chemistry have extended the versatility and flexibility of carbon–carbon bond formation in water or aqueous media. This chapter has highlighted the substantial progress that has been made in the last decade to 'tame' the reactive free radical species in aqueous phase reactions. The simplicity of water, its abundance and its wide availability will undoubtedly lead to the development of novel and exciting methodologies. The development of environmentally benign applications based on radical carbon–carbon bond formations is just around the corner.

References

1. B. Giese, *Radicals in Organic Chemistry Synthesis: Formation of Carbon–Carbon Bonds*, Pergamon Press, Oxford, 1986, and references cited therein.
2. D. P. Curran, in *Comprehensive Organic Synthesis*, eds. B. M. Trost, I. Fleming and M. F. Semmelheck, Pergamon Press, Oxford, 1991, vol. 4, p. 715.
3. H. Yorimitsu, K. Wakabayashi, H. Shinokubo and K. Oshima, *Tetrahedron Lett.*, 1999, **40**, 519; S Mikami, K. Fujita, T. Nakamura, H. Yorimitsu, H. Shinokubo, S. Matsubara and K. Oshima, *Org. Lett.*, 2001, **3**, 1853.
4. H. Yorimitsu, T. Nakamura, H. Shinokubo, K. Oshima, K. Omoto and H. Fujimoto, *J. Am. Chem. Soc.*, 2000, **122**, 11041.
5. K. Wakabayashi, H. Yorimitsu, H. Shinokubo and K. Oshima, *Bull. Chem. Soc. Jpn.*, 2000, **73**, 2377.
6. H. Yorimitsu, T. Nakamura, H. Shikubo, K. J. H. Oshima and K. Oshima, *Synlett*, 1998, **1351**.
7. K. Wakabayashi, H. Yorimitsu, H. Shinokubo and K. Oshima, *Bull. Chem. Soc. Jpn.*2000, **73**, 2377.
8. For reviews see: T. Naito, *Heterocycles*, 1999, **50**, 505; A.G. Fallis and I. M. Brinza, *Tetrahedron*, 1997, **53**, 17543; for intramolecular radical additions to C=N bonds: H. Migabe, O. Miyata and T. Naito, *J. Synth. Org. Chem. Jpn.*, 2002, **60**, 1087; H. Ishibashi, T. Sato and I. Ikeda, *Synthesis*, 2002, 695; G. K. Friestand, *Tetrahedron*, 2001, **57**, 5461; T. Naito, *Heterocycles*, 1999, **50**, 505; A. G. Fallis and I. M. Brinza, *Tetrahedron*, 1997, **53**, 17543; H. Miyabe, C. Ushiro, M. Ueda, K. Yamakawa and T. Naito, *J. Org. Chem.*, 2000, **65**, 176, and references cited therein.
9. H. Miyabe, M. Ueda, K. Fujii, A. Nishimura and T. Naito, *J. Org. Chem.*, 2003, **68**, 5618.
10. H. Miyabe and T. Naito, *Org. Biomol. Chem.*, 2004, 2, 1267, and references cited therein; B. C. Ranu and T. Mandal, *Tetrahedron Lett.*, 2006, 47, 2859; C.-J. Li and T. H. Chan, *Tetrahedron Lett.*, 1991, 32, 7017.
11. M. Ueda, H. Miyabe, A. Nishimura, O. Miyata, Y. Takemoto and T. Naito, *Org. Lett.*, 2003, **5**, 3835.

12. C.-J. Li, T.-H. Chan, *Tetrahedron Lett.*, 1991, **32**, 7017; C.-J. Li, *Chem. Rev.*, 1993, **93**, 2023, and references cited therein; V. T. Perchyonok, I. N. Lykakis and K. L. Tuck, *Green Chem.*, 2008, **10**, 153; V. T. Perchyonok and I. N. Lykakis, *Mini Rev. Org. Chem.*, 2008, **5**, 19; for some examples, see: I. H. S. Estevam and L. W. Bieber, *Tetrahedron Lett.*, 2003, **44**, 667; F. Marquez, R. Montoro, A. Liebaria, E. Lago, E. Molins and A. Delgado, *J. Org. Chem.*, 2002, **67**, 308.

13. J. S. Yadav and C. Srinivas *Tetrahedron Lett.*, 2002, **43**, 3837; A. Jeevanandam and Y. C. Ling *Tetrahedron Lett.*, 2001, **42**, 4361; B. Lavilla, O. Coll, J. Bosch, M. Orozco and F. J. Luque, *Eur. J. Org. Chem.*, 2001, 3719; D. Backhaus *Tetrahedron Lett.*, 2000, **41**, 2087; W. Lu and T. H. Chan, *J. Org. Chem.*, 2000, **65**, 8589; S. Hanessian, P.-P. Lu, J.-Y. Sanceau, P. Chemla, K. Gohda, R. Fonne-Pfister, L. Prade and S. W. Cowan-Jacob, *Angew. Chem. Int. Ed.*, 1999, **38**, 3160.

14. H. Miyabe, M. Ueda and T. Naito, *Chem. Commun.*, 2000, 2059; H. Miyabe, M. Ueda, A. Nishimura and T. Naito, *Org. Lett.*, 2002, **4**, 131.

15. C. Walling, *Tetrahedron*, 1985, **19**, 3887.

16. Y. Kita and M. Matsugi, in *Radical in Organic Synthesis*, eds. P. Renaud and M. P. Sibi, Wiley-VCH, Weinheim, 2001, vol. 1, pp. 1–10; for reviews, see: G. K. Friestad, *Tetrahedron*, 2001, **57**, 5461; H. Miyabe and T. Naito, *J. Synth. Org. Chem. Jpn.*, 2001, **59**, 35.

17. M. Ueda, H. Miyabe, A. Nishimura, O. Miyata, Y. Takemoto and T. Naito, *Org. Lett.*, 2003, **5**, 3835; H. Miyabe, M. Ueda and T. Naito, *J. Org. Chem.*, 2000, **65**, 5043.

18. H. Miyabe, M. Ueda, A. Nishimura and T. Naito, *Org. Lett.*, 2002, **4**, 131; H. Miyabe, M. Ueda, K. Fujii, A. Nishimura and T. Naito, *J. Org. Chem.*, 2003, **68**, 5618.

19. H. Miyabe and T. Naito, *Org. Biomol. Chem.*, 2004, **2**, 1267, and references cited therein; B. C. Ranu and T. Mandal, *Tetrahedron Lett.*, 2006, **47**, 2859.

20. G. R. Cook, S. Erickson and M. Hvinden, in *221st ACS National Meeting, San Diego, 1–5 April 2001*, vol. 4, p. 609.

21. D. O. Jang and D. H. Cho, *Synlett*, 2002, **631**.

22. R. Yanada, N. Nishimori, A. Matsumura, N. Fujii and Y. Takemoto, *Tetrahedron Lett.*, 2002, **43**, 4585.

23. T. Huang, C. C. K. Keh and C. -J. Li, *Chem. Commun.*, 2002, 2440.

24. M. Sugi, D. Sakuma and H. Togo, *J. Org. Chem.*, 2003, **68**, 7629.

25. H. Miyabe, M. Ueda and T. Naito, *Chem. Commun.*, 2000, 2059.

26. D. O. Jang and D. H. Cho, *Synlett.*, 2002, 631.

27. H. Miyabe, A. Nishimura, M. Ueda and T. Naito, *Chem. Commun.*, 2002, 1454; H. Miyabe, K. Fujii, T. Goto and T. Naito, *Org. Lett.*, 2000, **2**, 4071.

28. P. Renauld and M. Gerster, *Angew. Chem. Int. Ed.*, 1998, **37**, 2563; A. Ohno, M. Ikeguchi, T. Kimura and S. Oka, *J. Am. Chem. Soc.*, 1979, **101**, 7036; D. Nanni and D.P. Curran, *Tetrahedron: Asymmetry*, 1996, **7**, 2417; M. Blumenstein, K. Schwarzkopf and J. O. Metzger, *Angew. Chem. Int. Ed. Engl.*, 1997, **36**, 235; K. Schwarzkopf, M. Blumenstein, A. Hayen and J. O. Metzger, *Eur. J. Org. Chem.*, 1998, 177; A. J. Buckmelter, A. I.

Kim and S. D. Rychnovsky, *J. Am. Chem. Soc.*, 2000, 122, 9386; M. Helliwell, E. J. Thomas and L. A. Townsend, *J. Chem. Soc., Perkin Trans. 1*, 2002, 1286; C. H. Schiesser, *Arkivoc*, 2001, **2**, U34; D. Dakternieks and C. H. Schiesser, *Aust. J. Chem.*, 2001, **54**, 89; D. Dakternieks, K. Dunn, V. T. Perchyonok and C. H. Schiesser, *Chem. Commun.*, 1999, 1665; D. Dakternieks, K. Dunn, C. H. Schiesser and E. R. T. Tiekink, *J. Organomet. Chem.*, 2000, **605**, 209; V. T. Perchyonok and C. H. Schiesser, *Phosphorus Sulfur Silicon Relat. Elem.*, 1999, **151**, 193; M. A. Skidmore and C. H. Schiesser, *Phosphorus Sulfur Silicon Relat. Elem.*, 1999, **151**, 177; C. H. Schiesser, M. A. Skidmore and J. M. White, *Aust. J. Chem.*, 2001, **54**, 199; D. D. Tanner and A. Kharrat, *J. Am. Chem. Soc.*, 1988, **110**, 2968; H. Mokhtar-Jamai and M. Gielen, *Bull. Chem. Soc. Belg.*, 1975, **84**, 197; J. Vigeron and J. Jacquet, *Tetrahedron Lett.*, 1976, **32**, 939; M. Gielen and Y. J. Tondeur, *Organomet. Chem.*, 1979, **169**, 265; M. Ueda, H. Miyabe, A. Nishimura, H. Sugino and T. Naito, *Tetrahedron: Asymmetry*, 2003, **14**, 2857.

29. N. A. Porter, B. Giese and D. P. Curran, *Acc. Chem. Res.*, 1991, **24**, 296.
30. J. K. Kochi, *Free Radicals*, Wiley, New York, 1977.
31. D. J. Cram and F. A. Abd Ehafez, *J. Am. Chem. Soc.*, 1952, **74**, 5828; M. Cherest, H. Felkin and N. Prudent, *Tetrahedron Lett.*, 1986, 2199.
32. N. T. Ahn and O. Einstein, *Nouv. J. Chim.*, 1977, **1**, 61.
33. B. Giese, W. Damm, J. Dickhout, F. Wetterich, S. Sun and D. P. Curran, *Tetrahedron Lett.*, 1991, **32**, 6097.
34. S. Sun and D. P. Curran, *Tetrahedron Lett.*, 1993, **34**, 6181–6184.
35. D. P. Curran, N. A. Porter, B. Giese, *Stereochemistry of Radical Reactions*, VCH, Weinheim, 1995.

CHAPTER 8

Redox Processes and Electron Transfer: Free Radicals Are a Central Player

8.1 Electron Transfer Processes in General

In essentially all the reactions discussed so far, the radicals were generated by thermal or photochemical homolysis. There is another way to produce radicals and this is represented in Scheme 8.1. It consists in removing an electron from an electron-rich species represented by an anion or by adding one electron to an electron-deficient entity, now represented by a cation. It is possible, of course, to oxidise a radical to a cation or reduce it to an anion. This constitutes an alternative way of destroying radical character, in addition to recombination and disproportionation. Such transformations are referred to as redox processes; they are exceedingly important in radical chemistry and their impact on organic synthesis can hardly be overstated.

Generation of radicals by oxidation involves removing one electron from an anionic species or from a neutral electron-rich substance such as an enol. An electron is taken away from the π-orbital (the HOMO) to give a radical cation, which then loses a proton to give a free radical. As for producing radicals by reduction, addition of an electron to the antibonding orbital of R–X gives a radical anion, which rapidly collapses into a free radical $R^{\bullet}$ and an anion X^{-} (Scheme 8.2).

The rate of electron transfer reactions depends on the difference in the reduction potentials of educts and products. Since alkyl radicals possess an unpaired electron in a non-bonding orbital, electron transfer reactions to many metal salts often occur with high rates.[1] The higher are the SOMO energies of the radicals, the faster is the electron transfer.

RSC Green Chemistry No. 6
Radical Reactions in Aqueous Media
By V. Tamara Perchyonok
© V. Tamara Perchyonok 2010
Published by the Royal Society of Chemistry, www.rsc.org

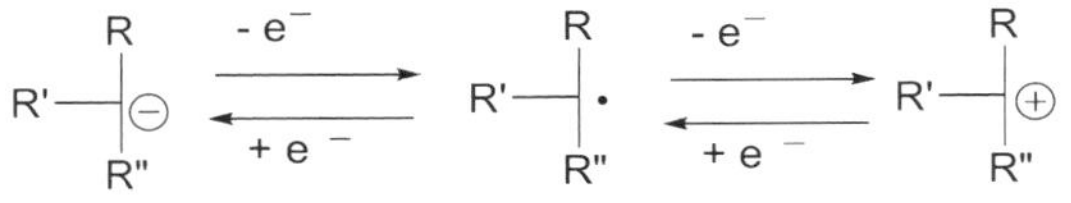

Scheme 8.1   Generation of radicals by electron transfer.

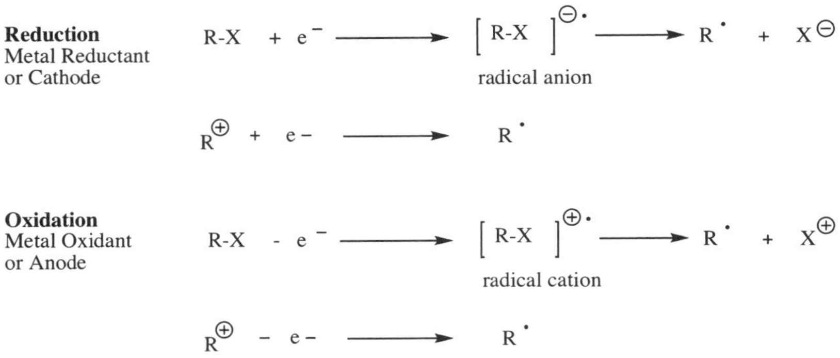

Scheme 8.2   Generation of radicals by reduction and oxidation.

8.2   Formation of Radicals by Oxidation with Transition Metal Salts: General Aspects and Considerations

Transition metals in a high oxidation state are often capable of extracting an electron from electron-rich organic substances. Ketones, esters, nitriles and various other 'carbon acids' that can form enols, enolates and related structures are by far the most commonly used substrates. Their oxidation can lead to a free radical, which then follows one or more of the pathways deployed in Scheme 8.3. Its important to take into account that the rate of radical production will depend on the exact structure of the substrate, its propensity to exist as the corresponding enol or enolate in the medium, the pH, the solvent, the temperature and, of course, the redox potential of the metallic salt (which can be strongly affected by the nature of the ligand around the metal) and the exact mechanism by which electron transfer actually occurs (*i.e.* inner or outer sphere)

The metal salts that are most commonly employed include Mn(III), Cu(II), Fe(III), Ce(IV), Ag(II) and Pb(IV). We will examine Mn(III) and Fe(III) in detail in turn, trying as far as possible to relate it back to Scheme 8.3.

8.2.1   Oxidations Involving Mn(III) in Aqueous, Ionic Liquids and the Solid State

Radical generation through oxidation of enolizable substrates using Mn(III) salts is by far the most common and the field is rapidly expanding. $Mn(OAc)_3 \cdot xH_2O$

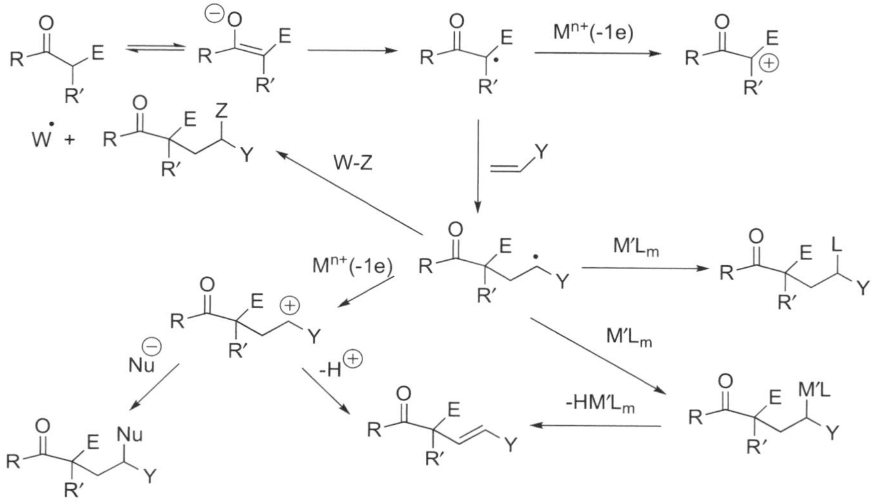

Scheme 8.3 Oxidation free radical transformations.

is the usual oxidant, which is actually a trimer made up of an oxo-centred triangle of Mn(III) ions bridged by acetate units; however, for simplicity and convenience, we shall use the simplified formula $Mn(OAc)_3$ throughout the rest of the examples. These reactions are considered to proceed via a free radical process in which Mn(III) initiates the oxidation of the carbonyl compound and then the newly formed radical undergoes an intermolecular addition to the alkenes to produce a new radical. The intramolecular addition of a radical to an aromatic ring gives rise to a stabilised radical, which is then oxidised with Mn(III) to restore the aromaticity and yield the corresponding furan.

Many of the methods that were previously employed for $Mn(OAc)_3$-mediated radical reactions involved the use of acetic acid as a solvent. Because of the poor solubility of $Mn(OAc)_3$ in organic solvents and the need for high temperatures for many reactions, the use of acetic acid limited the range of substrates that could be employed. In order to overcome this drawback, Parson investigated an elegant way of using ionic liquids to establish milder reaction conditions in $Mn(OAc)_3$-mediated reactions.[2] It was shown that ionic liquids, such as 1-butyl-3-methylimidazolium tetrafluoroborate ([bmim][BF$_4$]), which is miscible with polar solvents (*e.g.* methanol, dichloromethane) could be used in $Mn(OAc)_3$-mediated radical reactions.

The use of water as a solvent for this type of transformation is still in its infancy; however, Rickards described the efficient $Mn(OAc)_3$-mediated oxidative cyclisation of aryl-dicarbonyl and *Z*-unsaturated aryl-dicarbonyl compounds carrying electron-releasing groups in the aromatic ring, to afford 6,7,8,9-tetrahydro-5*H*-benzocycloocten-6-one and naphthalene-2(1*H*)-one, respectively (Scheme 8.4).[3]

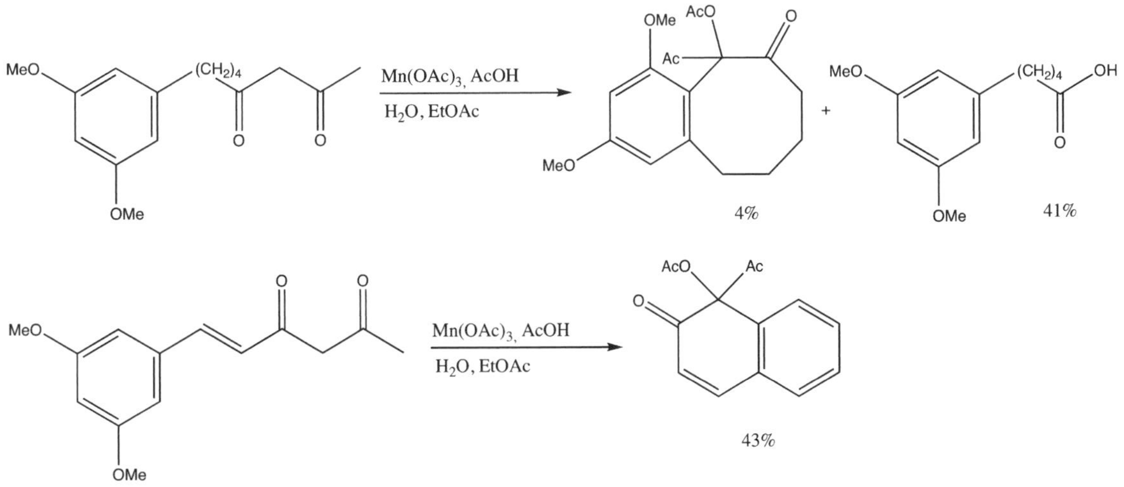

Scheme 8.4 Free radical cyclisation of 1,3-dicarbonyl compounds mediated by Mn(III) acetate.

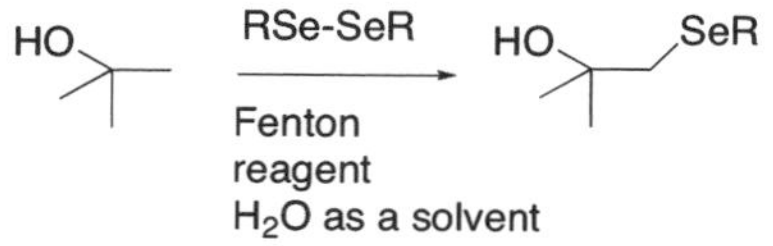

Scheme 8.5   Trapping of alkyl radicals in aqueous media using Fenton reagent.

8.2.2   Oxidations Using Fe(III) Salts (Fenton Reaction) in Aqueous Media: Fenton Chemistry is Yet Another Example of a Novel and Efficient Alkyl Radical Trap in Aqueous Medium

Fe(III) salts are known to oxidise electron-rich centres to foster the formation of radical species. They are particularly efficient in the oxidation of aromatic systems or a carbanion to the corresponding carbon-centred radical which undergoes C–C bond formation to yield the coupled products. For a successful synthesis, it is important to work in the absence of reactive synthetic molecules other than those which form the combination of radicals. Barton *et al.* used a simple water-soluble diselenide derivative that shows radical scavenger properties towards alkyl and hydroxyl radicals in Fenton-type chemistry (Fe^{2+}–H_2O_2).[4] The reaction rate between the produced alkyl radical and the diselenide overwhelms self-termination and halogen transfer reactions. The ability of diselenide to scavenge alkyl and hydroxyl radicals [$k_3(0\,^{\circ}C) = 6.1 \times 10^8\,M^{-1}\,s^{-1}$] could be exploited as a new tool in both synthetic and mechanistic work conducted in aqueous media (Scheme 8.5).[4]

8.3   Formation of Radicals by Reduction with Transition Metal Salts

8.3.1   General Reaction Considerations

The generation of radicals by electron transfer from low-valent transition metals is perhaps of even greater synthetic potential than that by oxidation of an electron-rich substrate, because a wider variety of precursors, reducing systems and radical traps may be used. In some cases, the reductant is sufficiently mild to be compatible with the presence of an oxidising agent and one or more oxidation steps can be incorporated into the sequence.

8.3.2   Reductions with Ti(III) Salts in Aqueous Media

Titanium(III) salts are strong reducing agents, capable of generating radicals by electron transfer from a variety of precursors. An example of the great advantage of using water as a hydrogen source and a useful synthetic medium is in the Cp_2TiCl-mediated rearrangement of 1,10-epoxy-11,13-dihydrocostunolide in which, for the first time, a hydrogen atom transfer from water to a carbon-centred radical was reported, suggesting that whatever the mechanism involved might turn out to be, its clear that in Ti(III)-mediated free radical chemistry water can act in a reductive way, working as a hydrogen atom donor. It is speculated that the generally accepted inertness of water in free radical chemistry should be carefully revised, especially in the presence of Ti(III) and other metal-centred free radicals (Scheme 8.6).[5]

The reactivity of water with both carbanion and carbocation intermediates is well known and recognised, but until recently it was generally believed that water is inert towards free radicals. Some years ago, Cuerva *et al.* by chance observed that tertiary radicals were reduced effectively in the presence of bis(cyclopentadienyl)titanium(III) chloride and water. Now the authors have solid evidence to show that water really acts as a complete hydrogen atom source rather than a simple proton donor for radical reductions mediated by Ti(III) and, presumably, other metals that react by single electron transfer (Scheme 8.7).[6]

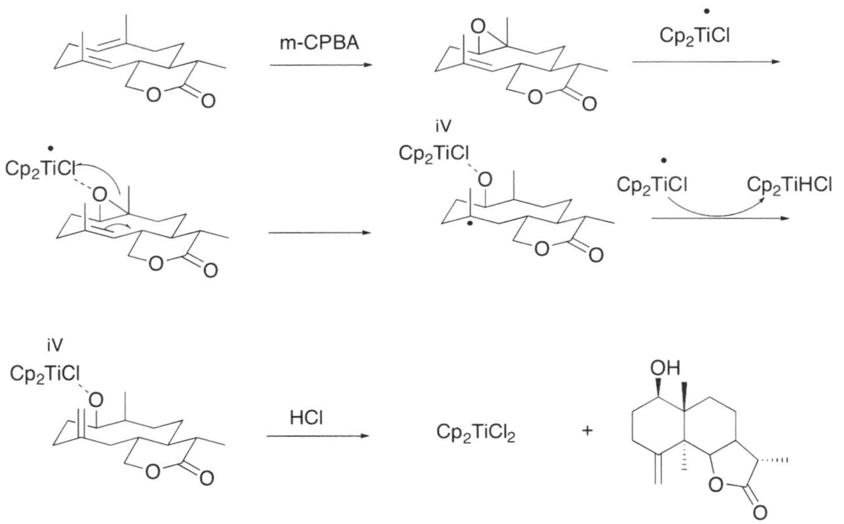

Scheme 8.6 Cp$_2$TiCl induced the deoxygenation and reduction of epoxide ring opening by radical chemistry.

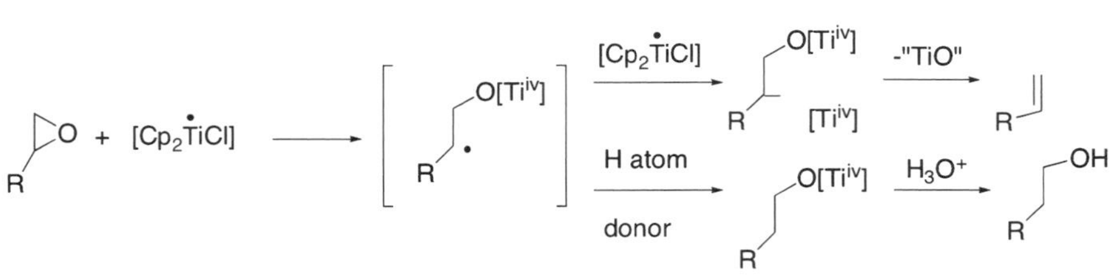

Scheme 8.7 Proposed mechanism of Ti-induced deoxygenation and reductive opening of epoxides.

8.3.3 Reduction with Sm(II) Iodide in Aqueous Media

Perhaps the most popular and synthetically versatile one-electron reducing agent is samarium diiodide, SmI_2. Its reducing properties were explored by Kagan's group and is a powerful reducing agent, capable of interacting usefully with a large variety of functional groups (Scheme 8.8).[7] Since then, two review papers on SmI_2-mediated cyclisations in organic chemistry have appeared.[8]

A useful solution for promoting difficult intermolecular acyl radical additions in the absence of a CO atmosphere, even in cases where the decarbonylation rate constant exceeds $10^8\,s^{-1}$, has been developed. The approach takes advantage of the ability of the single electron transfer agent samarium diiodide to reduce appropriately functionalised acyl derivatives to their corresponding ketyl radical anions, which subsequently react with acrylamides, acrylates or acrylonitrile, providing products arising from a formal acyl addition.[9]

The coupling protocol proved successful for all the *N*-acyl derivatives tested in aqueous media, even if the decarbonylation processes were too fast, as shown in Scheme 8.9.

SmI_2-mediated reductions of ketones, imines and α,β-unsaturated esters were reported by Hilmersson and co-workers to be instantaneous in the presence of H_2O and an amine in THF.[10] Not only were the SmI_2-mediated reductions shown to be fast and quantitative by the addition of H_2O and an amine, but also the workup procedures were simplified. Competing experiments with SmI_2–H_2O–amine confirmed that α,β-unsaturated esters could be selectively reduced in the presence of ketones or imines, as shown in Scheme 8.10.

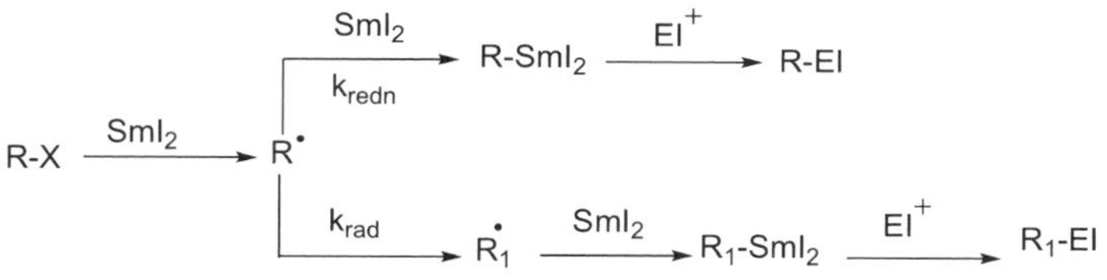
SmI2
R-SmI2
El+
R-El
kredn
SmI2
R-X
R•
krad
R1•
SmI2
R1-SmI2
El+
R1-El

Scheme 8.8 The reductive reaction conditions required for the generation of the radical in competition with the corresponding anion.

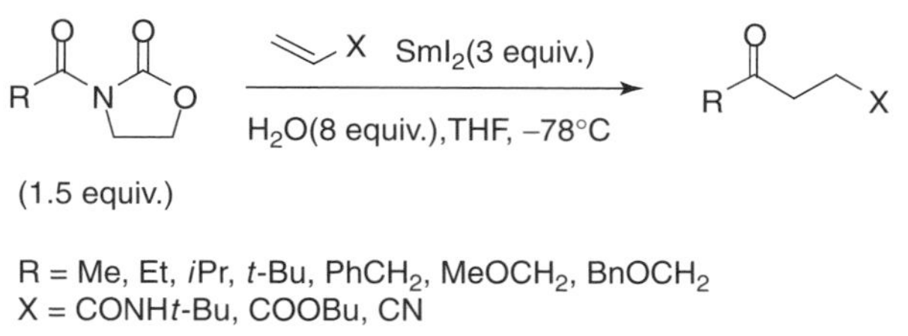
R
N
O
X SmI2(3 equiv.)
H2O(8 equiv.),THF, –78°C
R
X
(1.5 equiv.)
R = Me, Et, iPr, t-Bu, PhCH2, MeOCH2, BnOCH2
X = CONHt-Bu, COOBu, CN

Scheme 8.9 Radical addition of *N*-acyloxazolidinones to acrylamides and acrylates promoted by SmI_2–H_2O.

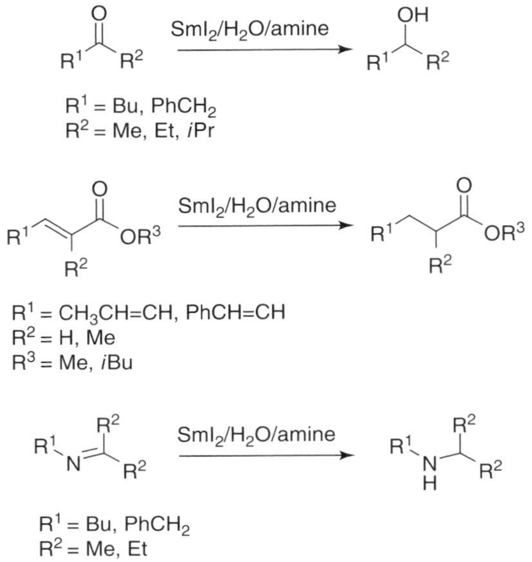

Scheme 8.10 SmI$_2$–H$_2$O–amine-mediated reduction of ketones, unsaturated esters and imines.

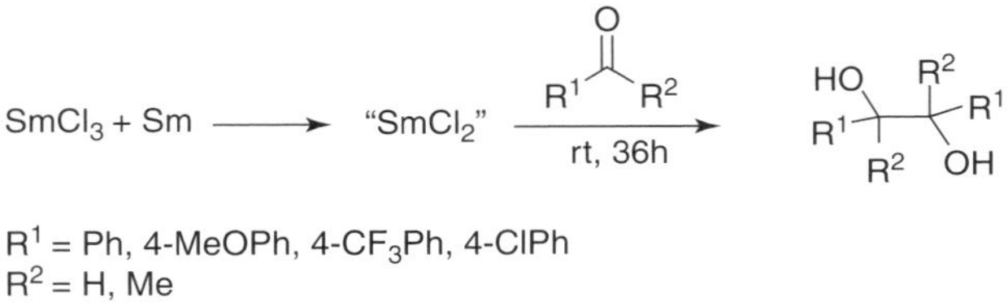

Scheme 8.11 Sm(II)-mediated pinacol coupling in water.

Comparison of analogous ligands showed that nitrogen and phosphorus ligands are superior to oxygen and sulfur ligands in these reductions.

In contrast, Matsukawa and Hinakubo proposed Sm(II)-mediated pinacol coupling in water (Scheme 8.11).[11] In all cases, the corresponding reduced product benzyl alcohols were formed in low yields. Unexpected disproportionation in water was also observed *via* UV–visible spectroscopic analysis. This indicated that low-valent samarium species can exist in water. Furthermore, the SmCl$_3$–Sm and SmCl$_3$–Mg systems were found to act as good one-electron reducing agents in water (Scheme 8.11).

Unlike aromatic carbonyl compounds, the reactions of aliphatic aldehydes and ketones using the Sm(II)–HCl–THF catalytic system gave exclusively the corresponding alcohols in high yields (Scheme 8.12).[12] In all cases, the corresponding pinacol products were formed in very low yields. In the case of *tert*-butyl

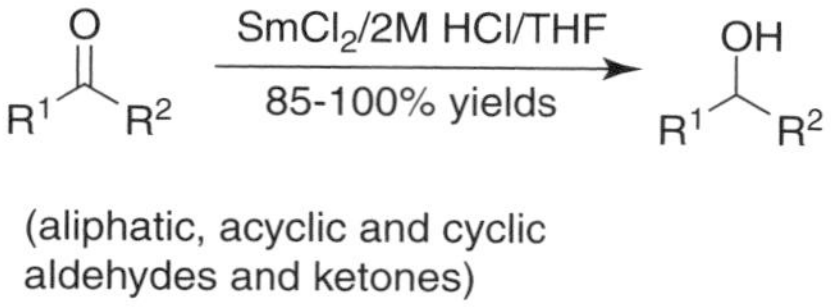

(aliphatic, acyclic and cyclic
aldehydes and ketones)

Scheme 8.12 Sm(II)-mediated reduction of aldehydes and ketones in water in the presence of additives.

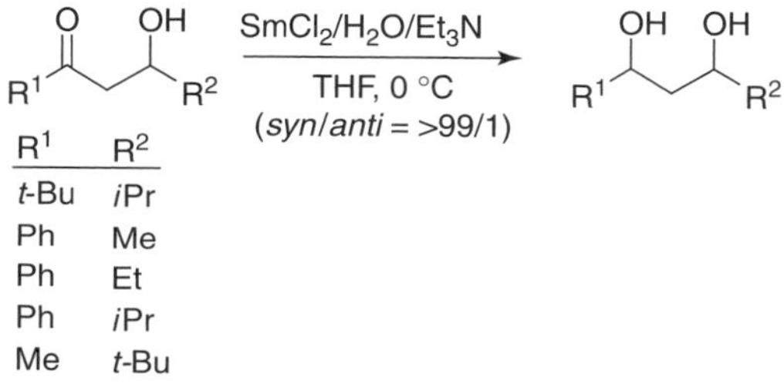

R^1	R^2
t-Bu	iPr
Ph	Me
Ph	Et
Ph	iPr
Me	t-Bu

Scheme 8.13 SmI$_2$–H$_2$O–Et$_3$N-mediated selective reduction of β-hydroxyketones to the corresponding 1,3-diols.

acetoacetate, the corresponding *tert*-butyl 3-hydroxybutanoate was formed in 97% yield without hydrolysis or reduction of the ester group. Reductions with SmI$_2$ are often conducted with an additive,[13] and the additive usually falls into one of the two classes: proton sources, as shown above, or electron donor molecules such as HMPA and DMPU.[13] The role of donor ligands was to increase the reducing power of Sm(II) and it has been suggested that proton sources served only to protonate the basic organometallic intermediates.[14] Under those conditions, DMPU was found to be an effective additive for the reduction of aryl iodide in acetonitrile but not THF.

Recently, Flowers and co-workers proposed the use of the SmI$_2$–H$_2$O–Et$_3$N system for the reduction of β-hydroxyketones to the corresponding 1,3-diols. In all cases, the reaction occurs without any by-product formation and with excellent diastereoselectivity with the *syn*-diol as the major isomer (Scheme 8.13).[15]

The SmI$_2$–H$_2$O–amine system has also been used for the selective cleavage of allyl ether-protected alcohols (Scheme 8.14). This method shown to be useful in the deprotection of alcohols and carbohydrates.[16]

During studies on SmI$_2$-mediated stereoselective spirocyclisation reactions, Procter and co-workers proposed the use of the SmI$_2$–H$_2$O catalytic system for the cyclisation of a methyl ketone to the corresponding spirocyclic lactone in 70% yield, as shown in Scheme 8.15.[17] Further treatment of the lactone with SmI$_2$–H$_2$O at room temperature gave the triol in high yield. Based on this result, they used also this catalytic system for the ring-selective reduction of a

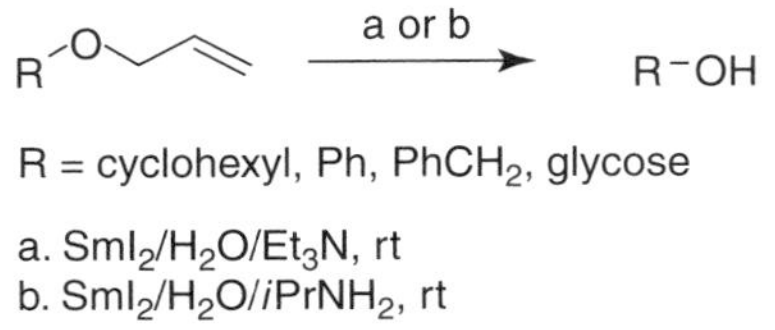

R−O−⟍⟍ $\xrightarrow{\text{a or b}}$ R−OH

R = cyclohexyl, Ph, PhCH$_2$, glycose

a. SmI$_2$/H$_2$O/Et$_3$N, rt
b. SmI$_2$/H$_2$O/iPrNH$_2$, rt

Scheme 8.14 SmI$_2$–H$_2$O–amine-mediated selective deprotection of allyl ethers to the corresponding alcohols.

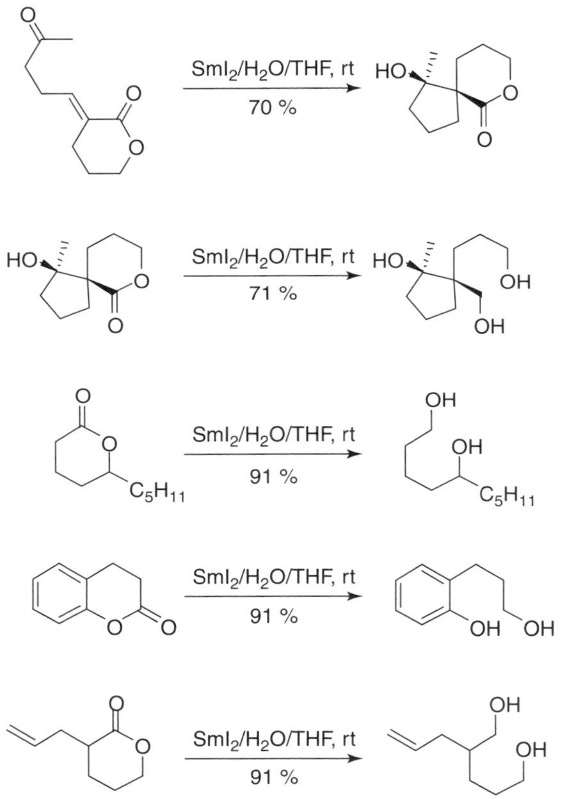

Scheme 8.15 Ring size-selective reduction of lactones using SmI$_2$ in H$_2$O.

variety of lactones, as shown in Scheme 8.15.[18] In all cases, the corresponding diols were formed in up to 85% yield.

Furthermore, it has been shown that the combination of amines and water has a remarkably strong effect on the reactivity of SmI$_2$ for the reduction of alkyl halides. In particular, alkyl chlorides are known to be more difficult to reduce than other functional groups, *e.g.* ketones, aldehydes and unsaturated esters. In

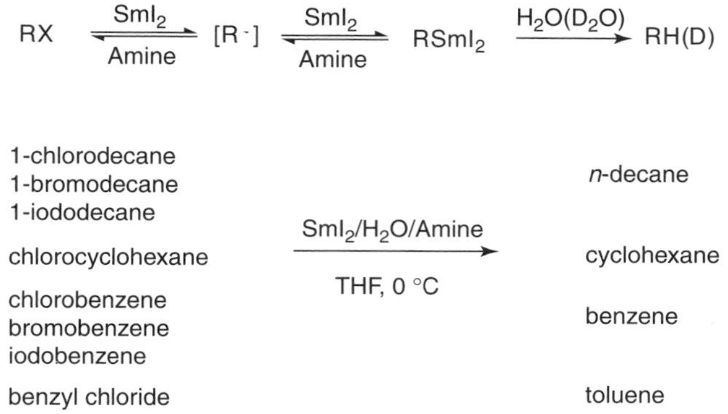

$$RX \underset{\text{Amine}}{\overset{SmI_2}{\rightleftharpoons}} [R^\cdot] \underset{\text{Amine}}{\overset{SmI_2}{\rightleftharpoons}} RSmI_2 \xrightarrow{H_2O(D_2O)} RH(D)$$

1-chlorodecane		*n*-decane
1-bromodecane		
1-iododecane	SmI$_2$/H$_2$O/Amine	
chlorocyclohexane	$\xrightarrow{\hspace{2cm}}$	cyclohexane
	THF, 0 °C	
chlorobenzene		benzene
bromobenzene		
iodobenzene		
benzyl chloride		toluene

Scheme 8.16 SmI$_2$–H$_2$O–amine-mediated reductions of alkyl and aromatic halides.

recent studies, SmI$_2$ was used to reduce alkyl halides to the corresponding alkanes (Scheme 8.16).[19] SmI$_2$–H$_2$O–amine mixtures have been found to reduce alkyl halides more efficiently than SmI$_2$–HMPA–alcohol mixtures at room temperature. Alkyl and aryl iodides were quantitatively reduced in <1 min and alkyl bromides in 10 min, whereas alkyl and aryl chlorides required more than 5 h for completion of the reaction. Determination of the reaction order of Et$_3$N in the reduction of 1-chlorodecane showed that the reaction order is one. Water was shown not to participate in the rate-determining step of this reduction.

The use of SmI$_2$ has been focused mainly on radical coupling reactions, but the discovery of SmI$_2$–water–amine mixtures has made 'SmI$_2$' a promising alternative to hydrides and hydrogen as a result of its mild but extremely fast reactions, and clean and simple workup procedures. The only requirement is that the reaction mixture is completely oxygen free. The mixture of samarium diiodide, water and amine (SmI$_2$–H$_2$O–Et$_3$N) is known to be a particularly powerful reductant, but until now the limiting reducing power has not been determined. A series of unsaturated hydrocarbons with varying half-wave reduction potentials (($E_{1/2}$) –1.6 to –3.4 V *vs.* SCE) were treated with SmI$_2$–H$_2$O–Et$_3$N and YbI$_2$–H$_2$O–Et$_3$N. All hydrocarbons with potentials of –2.8 V or more positive were readily reduced with SmI$_2$–H$_2$O–Et$_3$N, whereas all hydrocarbons with potentials of –2.3 V or more positive were readily reduced using YbI$_2$–H$_2$O–Et$_3$N. This defines limiting values of the chemical reducing power of SmI$_2$–H$_2$O–Et$_3$N as –2.8 V and of YbI$_2$–H$_2$O–Et$_3$N as –2.3 V *vs.* SCE.[20] Depending on the nature of the unsaturated hydrocarbon, a varying selectivity between different double bonds was observed, which gives rise to different product distributions. Reduction of polyaromatic hydrocarbons (PAHs) with fused rings, such as anthracene and phenanthrene, gave one product preferentially (Scheme 8.17), whereas aromatics with single bond-connected rings as in biphenyl and *p*-terphenyl gave a mixture of several isomeric products. For this reason, the synthetic use is limited to fused PAHs.

In addition to previous work, Zhan and co-workers confirmed water enhancement in SmI$_2$ catalytic reactions with the formation of alkanethiols

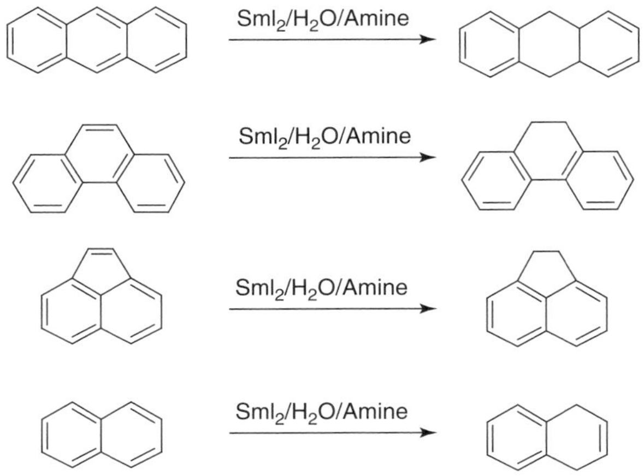

Scheme 8.17 SmI$_2$–H$_2$O–amine-mediated reductions of fused polyaromatic hydrocarbons.

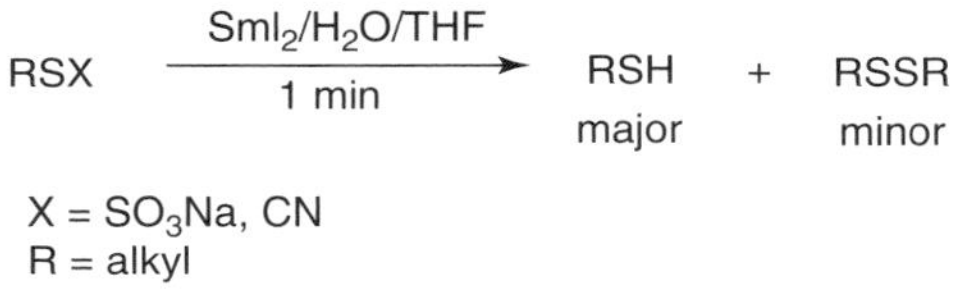

X = SO$_3$Na, CN
R = alkyl

Scheme 8.18 SmI$_2$–H$_2$O-mediated synthesis of alkanethiols from sodium alkyl thiosulfates and alkyl thiocyanates.

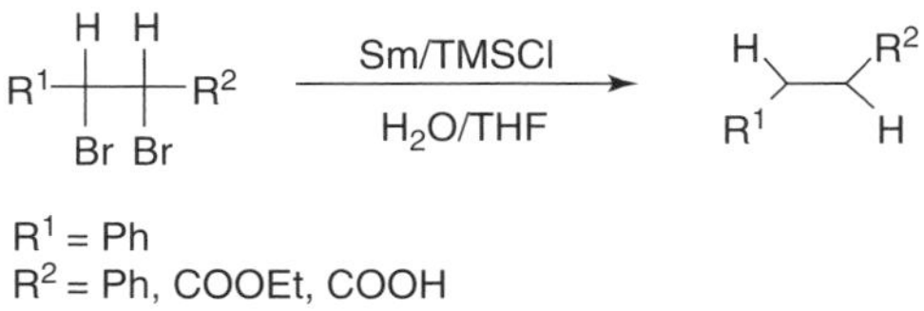

R^1 = Ph
R^2 = Ph, COOEt, COOH

Scheme 8.19 Dibromination of 1,2-dibromide derivatives mediated by Sm–TMSCl.

from the reduction of the corresponding alkyl thiocyanates and sodium alkyl thiosulfates (Scheme 8.18). The disulfide products were formed in very low relative yields, averaging from 1 to 19%.[21]

A simple and efficient method for the debromination of *vic*-dibromides to (*E*)-alkenes utilized the Sm–TMSCl–H$_2$O catalytic system (Scheme 8.19). For example, *trans*-stilbene was produced from 1,2-dibromo-1,2-diphenylethane within 5 h in good yield at room temperature. The benzylic *vic*-dibromides similarly gave the corresponding (*E*)-alkenes in a high yield. In the case of 1,2-dibromocyclohexane, a longer reaction time was needed to obtain cyclohexene. This is perhaps because the radical or anion intermediate of an aliphatic

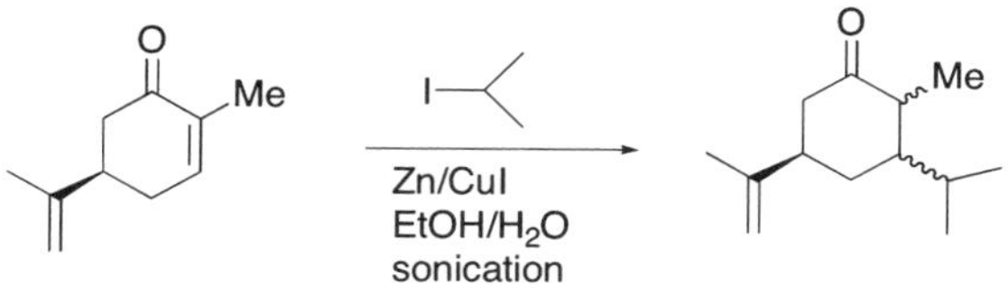

Scheme 8.20   Zinc-mediated conjugated addition of iodide.

vic-dibromide is less stable than that of the aromatic compound. Cholesterol can also he prepared from 5,6-dibromocholesterol under the same conditions.[22]

8.4   Dissolving Metal Reductions and Aqueous Media

Electron transfer from a dissolving metal also allows the generation and capture of radical intermediates. The reducing ability of the metal and the nature of the medium determine the type of functional groups that can be reduced. The limited number of reported reactions includes zinc metal, which, upon activation by sonication, is capable of producing radicals from aliphatic iodides, as shown in Scheme 8.20. Even though the reaction may also proceed by way of an organic zinc species, evidence has been provided that at least part of the pathway can be attributed to free radicals.[23,24]

8.5   Electron Transfer from Organic Reducing Agents in Aqueous Media

In a similar manner to low-valent and dissolving metals, electron-rich organic species can transport an electron to an appropriate substrate to give the corresponding radical anion, which then fragments in the usual fashion. The electron donor could be a neutral molecule or even a free radical. An important representative of the former case is tetrathiafulvalene (TTF) (Scheme 8.21).[24] TTF and its derivatives have often been used as components of organic metals because of their good electron donor ability and because they readily undergo reversible electron transfer. These features were elegantly exploited for the generation of aromatic radicals from diazonium salts. The transformation described, leading to various dihydrobenzofuran derivatives, provided a demonstration of the mechanistic versatility of this method, which allows a convenient and efficient crossover from the radical level to that of cation equivalent.

8.6   Conclusion

Although tremendous progress has been made in the application of redox processes for the generation and capture of various types of radicals, it is clear that many new areas remain to be explored. Numerous parameters can be

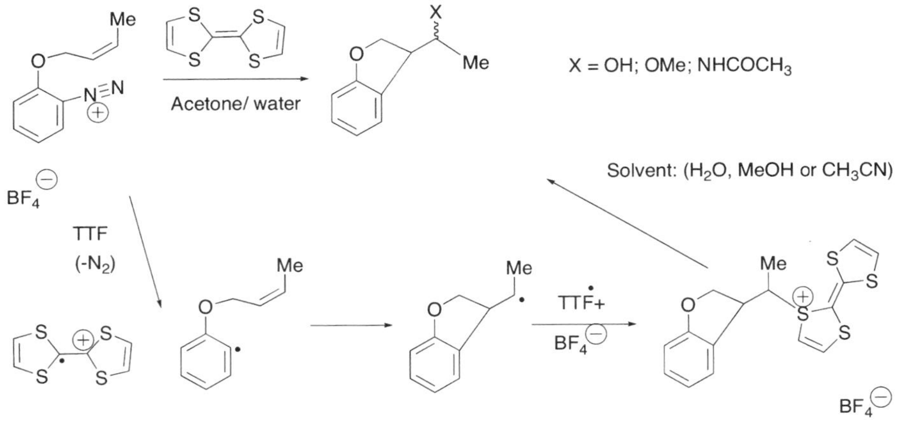

Scheme 8.21   Reduction of diazonium salts with tetrathiafulvalene.

modified in order to accomplish the desired transformation, not least concerning the nature of the metal or combination of metal salts or complexes. Understanding the subtle interplay of the various factors in this vast field of research is full of challenges and unexpected outcomes, which provide limitless possibilities for the conversion of simple substrates to complex products through one-pot, multi-step synthesis.

References

1. J. F. J. Coelho and A. M. F. P. Silva, *J. Polym. Sci., Part A: Polym. Chem.*, 2006, **44**, 3001.
2. G. Bar, A. F. Parson and C. B. Thomas, *Chem. Commun.*, 2001, 1350.
3. J. F. Jamie and R. W. Rickards, *J. Chem. Soc., Perkin Trans. 1*, 1996, 2603; J. F. Jamie and R. W. Rickards, *J. Chem. Soc., Perkin Trans. 1*, 1997, 3613.
4. D. H. R. Barton, M. Jacob and E. Peralez, *Tetrahedron Lett.*, 1999, **40**(52), 9201.
5. A. J. Barrero, J. E. Oltra, J. M. Cuerva and A. Rosales, *J. Org. Chem.*, 2002, **67**, 2566; A. Rosales, J. M. Cuerva, J. E. Oltra, *Catalyst for the Fine Chemical Synthesis*, 2007, **5**, 97.
6. D. J. Cardenas, E. Bunuel and J. E. Oltra, *Angew. Chem. Int. Ed.*, 2006, **45**, 5522.
7. P. Girard, J. -L. Namy and H. B. Kagan, *J. Am. Chem. Soc.*, 1980, **102**, 2693.
8. H. B. Kagan, *Tetrahedron*, 2003, **59**, 10351; D. J. Edmonds, D. Johnston and D. J. Procter, *Chem. Rev.*, 2004, **104**, 3371.
9. C. M. Jensen, K. B. Lindsay, R. H. Taaning, J. Karaffa, A. M. Hansen and T. Skrydstrup, *J. Am. Chem. Soc.*, 2005, **127**, 6544.

10. A. Dahlen and G. Hilmersson, *Chem. Eur. J.*, 2003, **9**, 1123; A. Dahlen and G. Hilmersson, *Tetrahedron Lett.*, 2002, **43**, 7197.
11. S. Matsukawa and Y. Hinakubo, *Org. Lett.*, 2003, **5**, 1221–1223.
12. S. Talukdar and J. -M. Fang, *J. Org. Chem.*, 2001, **66**, 330.
13. P. Girard, J. -L. Namy and H. B. Kagan, *J. Am. Chem. Soc.*, 1980, **102**, 2693.
14. G. H. Molander, *Organic Reactions*, 1994, **46**, 211; G. A. Molander and C. R. Harris, *Chemical Reviews*, **96**, 307.
15. T. A. Davis, P. R. Chopade, G. Hilmersson and R. A. Flowers, II, *Org. Lett.*, 2005, 7, 119, and references cited therein.
16. A. Dahlen, A. Sundgren, M. Lahmann, S. Oscarson and G. Hilmersson, *Org. Lett.*, 2003, **5**, 4085.
17. T. K. Hutton, K. W. Muir and D. J. Procter, *Org. Lett.*, 2003, **5**, 4811.
18. L. A. Duffy, H. Matsubara and D. J. Procter, *J. Am. Chem. Soc.*, 2008, 130, 1136, and references cited therein.
19. A. Dahlen, G. Hilmersson, B. W. Knettle and R. A. Flowers II, *J. Org. Chem.*, 2003, **68**, 4870.
20. A. Dahlen, A. Nilsson and G. Hilmersson, *J. Org. Chem.*, 2006, **71**, 1576.
21. Z.-P. Zhan, K. Lang, F. Liu and L.-M. Hu, *Synth. Commun.*, 2004, **34**, 3203.
22. X. Lu, P. Lu and Y. Zhang, *Synth. Commun.*2000, **30**, 1917.
23. J.-L. Luche and C. Allavena, *Tetrahedron Lett.*, 1988, **29**, 5369;, J.-L. Luche, C. Allavena, C. Petrier and C. Dupuy, *Tetrahedron Lett.*, 1988, **29**, 5373; P. Blanchard, A. D. Da Silva, J.-L. Fourrey, A. S. Machado and M. Robert-Gero, *Tetrahedron Lett.*, 1992, **33**, 8069; P. Blanchard, M. S. El Kortbi, J.-L. Fourrey and M. Robert-Gero, *J. Org. Chem.*, 1993, **58**, 6517.
24. N. Bashir and J. A. Murphy, *Chem. Commun.*, 2000, 627.

CHAPTER 9

Chain versus Non-chain Free Radical Processes in Aqueous Media

9.1 The Persistent Radical Effect: General Principles and Considerations

In earlier chapters, the tremendous advantages of and achievements in the field of free radical chain reactions were highlighted, with the desired end product being predetermined in the propagation steps. The faster the propagation step, the greater is the observed efficiency: less initiator is needed, fewer unwanted side reactions can compete and radical-radical interactions constituting the termination become negligible. In summary, the whole approach was aimed at reducing radical–radical interactions by keeping the steady-state concentration of the intermediate radical species as low as possible. Since the interactions are usually diffusion controlled and therefore unselective, it might be considered (at first glance) an impossible task to gain control of an inter-radical reaction for synthetic purposes. However, this is not a case at all, due to the elegant and ingenious solution to this problem based on the persistent radical effect, also known as the Fischer–Ingold effect. This phenomenon, which has only recently been elucidated, underlies several reactions that occur in Nature and also novel synthetic applications recently discovered and elegantly applied in the special field of living radical polymerisation.

The Fischer–Ingold effect can be understood without the need for elaborate – but more rigorous – mathematical modelling. Let us consider a compound A, which can be decomposed thermally or photochemically (with or without solvent) into two radicals, $X^\bullet$ and $Y^\bullet$, which will then undergo recombination. If we neglected eventual cage effects and assume that the rate of such recombinations is diffusion controlled (and therefore comparable), then, statistically, of the three possible reactions shown by in Scheme 9.1, one would expect a yield

RSC Green Chemistry No. 6
Radical Reactions in Aqueous Media
By V. Tamara Perchyonok
© V. Tamara Perchyonok 2010
Published by the Royal Society of Chemistry, www.rsc.org

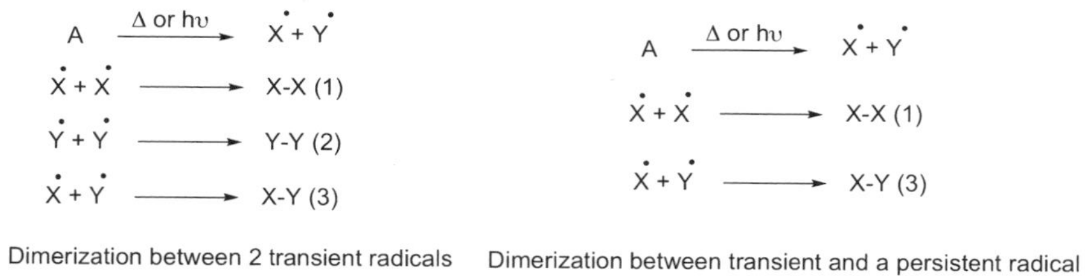

$A \xrightarrow{\Delta \text{ or } h\upsilon} \overset{\bullet}{X} + \overset{\bullet}{Y}$

$\overset{\bullet}{X} + \overset{\bullet}{X} \longrightarrow X\text{-}X \ (1)$

$\overset{\bullet}{Y} + \overset{\bullet}{Y} \longrightarrow Y\text{-}Y \ (2)$

$\overset{\bullet}{X} + \overset{\bullet}{Y} \longrightarrow X\text{-}Y \ (3)$

Dimerization between 2 transient radicals

$A \xrightarrow{\Delta \text{ or } h\upsilon} \overset{\bullet}{X} + \overset{\bullet}{Y}$

$\overset{\bullet}{X} + \overset{\bullet}{X} \longrightarrow X\text{-}X \ (1)$

$\overset{\bullet}{X} + \overset{\bullet}{Y} \longrightarrow X\text{-}Y \ (3)$

Dimerization between transient and a persistent radical

Scheme 9.1   Dimerization between two transient radicals (left) and a transient and a persistant radical (right)

of roughly 25% of each of the symmetrical dimmers X–X and Y–Y and a 50% yield of the cross-product X–Y. There is therefore already some statistical selectivity in favour of the latter.

Now, what will the effect on the selectivity be if $Y^{\bullet}$ is a persistent radical, that is, if reaction (2) does not occur? We are now left with two reactions, (1) and (3), as shown in Scheme 9.1, and the first answer which comes to mind is that the selectivity in favour of X–Y would increase to around 66%, again on statistical grounds. In fact, the selectivity for the cross-product becomes almost total. The reason is that we are not dealing with a hypothetical mathematical situation where all $X^{\bullet}$ and $Y^{\bullet}$ are generated instantly and react instantly.

In this real-life chemical transformation, compound A is being decomposed over a certain period (minutes, hours) which is vastly greater then the lifetime of the intermediate radicals (micro- to nanoseconds), whose concentration in the medium remains low throughout. Now, while the decomposition of A and the cross-coupling reaction (3) do not modify the relative concentration of $X^{\bullet}$ and $Y^{\bullet}$ (either one of each is produced or one of each is consumed), the formation of X–X in reaction (1) consumes in contrast only $X^{\bullet}$ radicals. The consequence is that very soon after the radicals start to be generated, the relative concentration becomes tilted greatly in favour of radicals $Y^{\bullet}$, even if the absolute, steady-state concentration of both $X^{\bullet}$ and $Y^{\bullet}$ remains small. Thus, every time a radical $X^{\bullet}$ is created by the decomposition of A, its chances of capturing a radical $Y^{\bullet}$ are much greater than that of capturing another $X^{\bullet}$; of course, radicals $Y^{\bullet}$ being by definition persistent can only react with $X^{\bullet}$. The formation of the cross-coupling product X–Y will therefore rapidly dominate and the selectivity, for all practical purposes, will be almost complete. The same result will be qualitatively obtained if reaction (2) is (comparatively) slow or easily reversible under the experimental conditions.

9.2   Living Free Radical Polymerisations in Aqueous Media

A rapidly emerging and blossoming area in radical chemistry is living free radical polymerisation. Notwithstanding their enormous importance, radical polymerisations in general are not within the scope of this book; nevertheless, since much of the recent work in the special field of living-radical polymerisation involves persistent nitroxyl radicals and the application of the Fischer–Ingold effect, a short description is useful.[1]

When a monomer such as a styrene or methyl acrylate is polymerised with a radical initiator, one obtains a polymer where the chain has varying length and the average molecular weight depends on (amongst other factors) the ratio of monomers to initiator. Since the polymerisation terminates irreversibly through dimerisation or disproportionation, the distribution of molecular weight is fairly large (large polydispersity) and it is not generally easy to build block copolymers where a segment made of a given monomer is followed by one or more segments made from other monomers. Also, even if such block macromolecules were constructed, the wide polydispersity in each of the constituent segments would translate into poor control over the mechanical and other properties of the end product. These difficulties can sometimes be overcome by employing living ionic polymerisation, but the technical hurdles on scale-up are challenging, with constraints with respect to very pure monomers and solvents, rigorous absence of water and oxygen and low compatibility with many functional groups.

Atom transfer radical additions and cyclisations have been used successfully in organic chemistry for the preparation of 1:1 adducts from alkyl halides, RX, and alkenes, $CH_2=CHY$ (Scheme 9.2) Under such conditions, the required catalytic amount of transition metal, M_t^n (*e.g.* CuCl, FeBr$_2$, RuCl$_2$ in the presence of corresponding ligand) is used to provide a low stationary concentration of radicals, $R^{\bullet}$ (and of oxidised transition metal $M_t^{n+1}X$, *e.g.* CuCl$_2$), which subsequently react with an alkene by abstraction of a halogen atom from the oxidised form of the catalyst to produce the final product, R–CH$_2$–CHY–X.

One of the most distinguishable features of radical polymerisation is its tolerance to water, relative to the ionic counterparts, however the effective polymerization should be performed conventionally under stringent conditions without protonic or basic impurities to insure effective chain propagation and therefore desired polymer growth without unnecessary inhibition and premature termination. Because of their unique features, suspension, dispersion

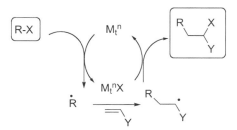

Scheme 9.2   Schematic example of the atom transfer radical polymerisation technique to prepare an adduct from an alkyl halide (R–X) and an alkene (CH$_2$ =CHY). M_t^n is the transition metal; M_t^{n+1} is the oxidised transition metal; X is a halogen and R is a radical.

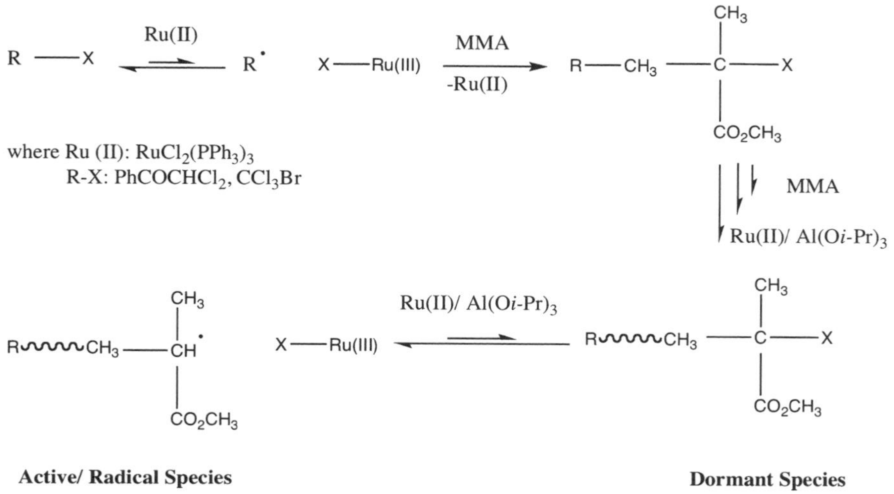

Active/ Radical Species   **Dormant Species**

Scheme 9.3   Living polymerisation in aqueous media.

and emulsion processes in aqueous and alcoholic media are widely employed in radical polymerisation. Sawamoto and co-workers developed living radical polymerisation of methyl methacrylate (MMA) and related acrylic and styrenic monomers mediated by a ruthenium complex $[RuCl_2(PPh_3)_3]$ in water and alcohols (suspension process).[2] The living polymerisation of interest utilised MMA in water and alcohol by using $[RuCl_2(PPh_3)_3]$ in conjunction with an organic halide initiator such as $PhCOCHCl_2$ or CCl_3Br (Scheme 9.3).

Coelho and co-workers have recently reported another use of water in living polymerisation in the synthesis of a block copolymer [poly(vinyl chloride)-b-poly(*n*-butyl acrylate)-b-poly(vinyl chloride)].[3] The new material was synthesised by single electron transfer/degenerative chain transfer-mediated living radical polymerisation (SET-DTLRP) in two steps.

9.3   Mimicking Vitamin B_{12}: the Radical Chemistry of Organocobalt Derivatives in Water and Aqueous Media

The cobalt-containing vitamin B_{12} or cyanocobalamin and its biologically active form, coenzyme B_{12} or adenosylcobalamin (Figure 9.1), represent some of the most complex organometallic compounds found in nature.

Coenzyme B_{12} mediates a number of important biochemical transformations where the first step appears to be the homolytic rupture of the carbon–cobalt bond of the adenosyl moiety to give a primary 5′-adenosyl radical.[4]

The rich and versatile chemistry of cobalamin has inspired a vast amount of work, both mechanistic and synthetic. The key features is the weakness of the C–Co bond (20–30 kcal mol^{-1}), which upon thermolysis or more commonly photolysis (visible light) gives a carbon radical and a persistent Co(II)-centred radical.

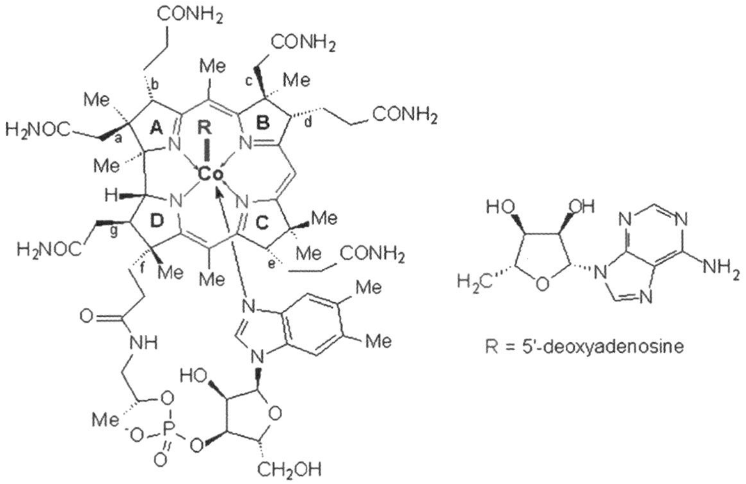

Figure 9.1 Vitamin B_{12}.

Fischer–Ingold control of selectivity operates, allowing numerous useful synthetic transformations.[5]

Vitamin B_{12} and its derivatives have attracted considerable attention for their use in synthetic organic chemistry, including several recent new applications.[6] Vitamin B_{12} contains a cobalt atom surrounded by an equatorial corrin ring system. The chiral environment of the corrin ring has been utilised for the development of stereoselective B_{12}-catalysed reactions.[7] Covalently linked to the corrin is a dimethylbenzimidazole that reversibly occupies one of the axial coordination sites.[7] Van der Donk and co-workers reported the use of vitamin B_{12} for a broad range of free radical synthetic transformations (*i.e.* intramolecular cyclisation reactions that utilise vitamin B_{12} as a catalyst and Ti(III) citrate as a reducing agent in aqueous media).[7] In addition, the reaction mechanism was explored, uncovering an interesting reversal of product distribution by altering the reaction pH (Scheme 9.4).[1,7]

According to this mechanism, the formation of the unsaturated product is dependent on β-hydrogen atom abstraction by Co(II)-cobalamin.[1,7] However, the concentration of the latter at any time is fairly low since it is rapidly reduced to the Co(I)-cobalamin by Ti(III) citrate. Hence its is speculated that by slowing this reduction step and using solvents without readily available hydrogen atoms, the product ratio could be significantly influenced. The reduction potential of Ti(III) citrate has been shown to decrease in a pH-dependent manner. A smaller thermodynamic driving force for reduction of Co(II) to Co(I) at lower pH could result in a kinetically slower reduction process, which might make the hydrogen atom abstraction more competitive.

The scope of the transformation is summarised in Scheme 9.5.

The novel cyclisation reactions are very attractive as they represent a mild, environmentally friendly process with a possible pH-controlled product

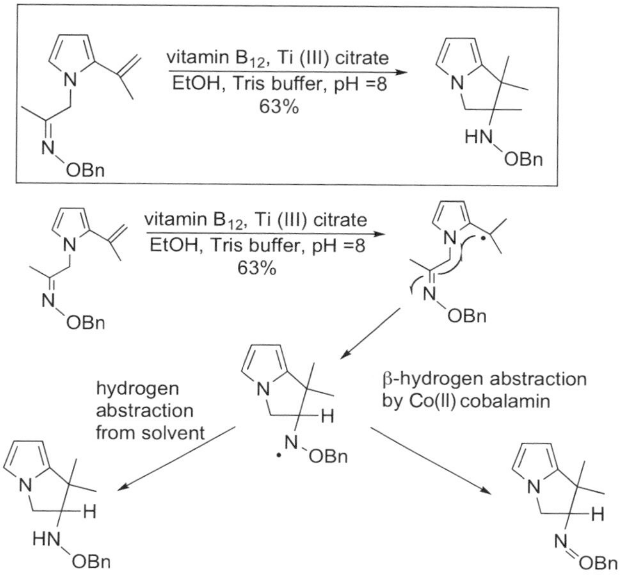

Scheme 9.4 Vitamin B_{12}-catalysed radical cyclisation of oxime and proposed catalytic mechanism for the reaction.

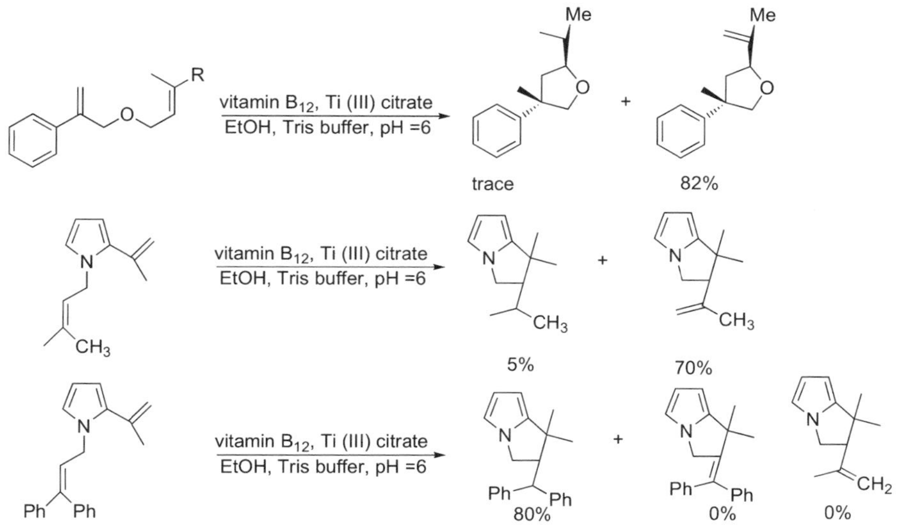

Scheme 9.5 Screening of substrates for intramolecular vitamin B_{12}-catalysed cyclisations at various pH values.

distribution. An added advantage of the methodology lies in the great degree of regio- and stereocontrol through the well-defined regiochemistry of the activation of the arylalkene, which is not readily available through other means.

References

1. A. Bravo, H.-R. Bjørsvik, F. Fontana, L. Liguori and F. Minisci, *J. Org. Chem.*, 1997, **62** (12), 3849–3857, and references cited therein; H. Fischer, Macromolecules, 1997, 30 (19), 5666–5672; A. Studer and S. Amrein, *Angew. Chem. Int. Ed.*, 2000, **39**, 3080.
2. M. Kato, M. Kamigaito, M. Sawamoto and T. Higashimura, *Macromolecules*, 1995, **28**, 1721–1723.
3. J. F. J. Coelho, A. M. F. P. Silva, A. V. Popov, V. Percec, M. V. Abreu, P. M. O. F. Goncalves and M. H. Gil, *J. Polym. Sci. Part A: Polym. Chem.*, 2006, **44**(9), 2809–2825.
4. P. A. Frey, Importance of organic radicals in enzymatic cleavages of unactivated C–H bonds, *Chem. Rev.*, 1990, **90**, 1343–1357.
5. U. E. Krone, R. K. Thauer and H. P. C. Hogenkamp, *Biochemistry*, 1989, **28** (11), 4908–4914; R. Scheffold, S. Albrecht, R. Orlinski, H.-R. Rut, P. Stamouli, O. Tinembart, L. Walder and C. Weymuth, *Pure Appl. Chem.*, 1987, **59**, 363–372.
6. K. L. Brown, *Chem. Rev.*, 2005, **105**, 2075.
7. J. Shey, C. M. McGinley, K. M. McCauley, A. Dearth, B. Young and W. A. van der Donk, *J. Org. Chem.*, 2002, **67**, 837; C. L. Forbes and R. W. Franck, *J. Org. Chem.*, 1999, **64**, 1424; M. Kleban, U. Kautz, J. Greul, P. Hilgers, R. Kugler, H.-Q. Dong, V. Jager, *Synthesis*, 2000, 1027; L. Huang, Y. Chen, G.-Y. Gao and X. P. Zhang, *J. Org. Chem.*, 2003, **68**, 8179; Y. Chen and X. P. Zhang, *J. Org. Chem.*, 2004, **69**, 2431.

CHAPTER 10

Future Directions and Practical Considerations

The current view of radical chemistry is slowly but surely establishing itself as a force to reckon with in the curriculum of undergraduate and postgraduate students. The earlier chapters gave only a brief taste of the spectrum of free radical chemistry available to the synthetic chemists to utilize and advance chemistry and understanding. Its envisaged that the chemistry described will only be limited by the creativity of the researcher.

As in other ways of life, practicing and making mistakes is still the best way to learn. Here we give some practical considerations that might make the process more fruitful and successful, based on a wealth of experience in free radical chemistry in conventional solvents.

- The presence of dissolved oxygen may be the complicating factor, especially when working on a small scale, below the boiling point of the solvent. One might need to look for an efficient degassing procedure.
- If the desired transformation is intramolecular processes, then the dilution factor might be worth considering, as increasing the dilution will give an advantage over the bimolecular processes. If the key step is a fragmentation, then implying positive entropy through the increase in temperature will have a beneficial effect in speeding up the fragmentation with respect to other competitive inter- or intramolecular pathways, which usually have a negative entropy. Sometimes the order of addition of reagents can also influence the outcome.
- One frequent observation is that radicals can more frequently be reduced instead of undergoing the usually more fancy but desired transformation. The unexpected hydrogen donor can be a reagent (such as a stannane, silane, thiol, hypophosphorous acid, *etc.*, or solvent). If the hydrogen shift is suspected to be a culprit, it could be easily unmasked through a deuterium experiment. It may then be necessary to block the unwanted

RSC Green Chemistry No. 6
Radical Reactions in Aqueous Media
By V. Tamara Perchyonok
© V. Tamara Perchyonok 2010
Published by the Royal Society of Chemistry, www.rsc.org

hydrogen shift by modifying the substituent/protecting group in order to alter the folding of the molecule and perhaps alter the internal migration of the hydrogen atom.

- Protecting or temporary groups may be used to accelerate or disfavour a given reaction. The whole range of polar and steric effects of substituents (including the Thorpe–Ingold effect) may be utilized to control the rate of the desired transformation. The temporary tether has been used to great effect in many situations to control the regio- and stereochemical outcomes of free radical transformations in organic media and lately in aqueous media.
- If a problem appears to be due to unwanted radical–radical interactions, then it might be necessary to slow the initiation processes. This can be achieved by adjusting the temperature of the reaction, the decomposition temperature of the initiator and the rate of introduction of the initiator. In case of photochemical initiation, the power of the lamp could have an influence or for redox processes, one of the reagents could be introduced more slowly.
- Any bottleneck in the reaction steps should be identified as early in the sequence as possible and appropriate structural modifications to the precursors/reagents could be introduced early on in the sequence through appropriate temporary substitutions.
- Choose the initiator which is appropriate to your system. AIBN, although capable of doing many things, is often incapable of triggering atom or group transfer reactions (try lauroyl peroxide instead).

As in anything in life, practice and making errors is the best way to learn and master the skills and techniques. Its hoped that this book will be an aid for organic chemists to become skilled practitioners in the exciting and challenging field of green free radical chemistry.

The book is concerned with bringing together the collective knowledge and principles of radical chemistry in aqueous media. Yet the broad range of applications of free radical chemistry in synthesis is a never-ending sense of wonder and surprise for traditional organic chemists.

10.1   Generation of Radicals for Probing DNA Structures

The various conformations of DNA – the A, B and Z forms – the protein-induced DNA kink and the G-quartet form, are thought to play important biological roles in processes such as DNA replication, gene expression and regulation and the repair of DNA damage.[1] The investigation of local DNA conformational changes associated with biological events is therefore essential for understanding the function of DNA. Damage to the sugar unit of DNA is mainly determined by the formation of free radicals through hydrogen atom abstraction from one of the five available positions. The radicals *in vivo* can repair themselves by hydrogen atom abstraction from glutathione or can lead to modification of the sugar unit (DNA damage) or can lead to strand breakage, and therefore the ability of chemists to generate the desired radicals under 'Nature-like conditions' is essential to move forward our understanding of radical-mediated DNA damage. There are several ways of generating C1', C2', C3', C4' and C5' radicals under 'aqueous conditions' which are summarized in Scheme 10.1.

10.2 Helpful Hints and Tips: Radical Generation at the Fingertips

This section gives a concise collection of the methods that lead to the generation of carbon-centred radicals *via* homolytic cleavage of C–X bonds and which are used in the formation of C–H or C–C bonds. Detailed discussions and examples of the transformations in detail are reported in earlier chapters.

10.2.1 Carbon–Halogen Bonds

Alkyl and aryl halides

$RX \rightarrow R^{\bullet} \longrightarrow$

- Bu_3SnH, AIBN or *hv*
- Bu_3GeH, AIBN
- TMS_3SiH, initiator
- H_3PO_2, initiator
- DEPO, initiator
- $InCl_3/Bu_3SnH$
- $InCl_3/ET_3SiH$
- $InCl_3/NaBH_4$
- $XCo(dmgH)_2py$, $NaBH_4$ or cathode
- $Co(dmgH)_2py$
- Vitamin B_{12}, Zn or cathode
- Na or K
- Anions
- Bz_2O_2, Fe^{3+}
- AIBN or *hv*

10.2.2 Carbon–Oxygen Bonds

Alcohols

$ROH \rightarrow R^{\bullet} \longrightarrow$

- CS_2, MeI, Bu_3SnH, AIBN
- CS_2, MeI, TMS_3SiH, initiator
- CS_2, MeI, H_3PO_2/additive, initiator
- $(COCl)_2$, *N*-hydroxypiperidine-2-thione, initiation
- $ClPO(OEt)_2$; $S_{RN}1$ conditions

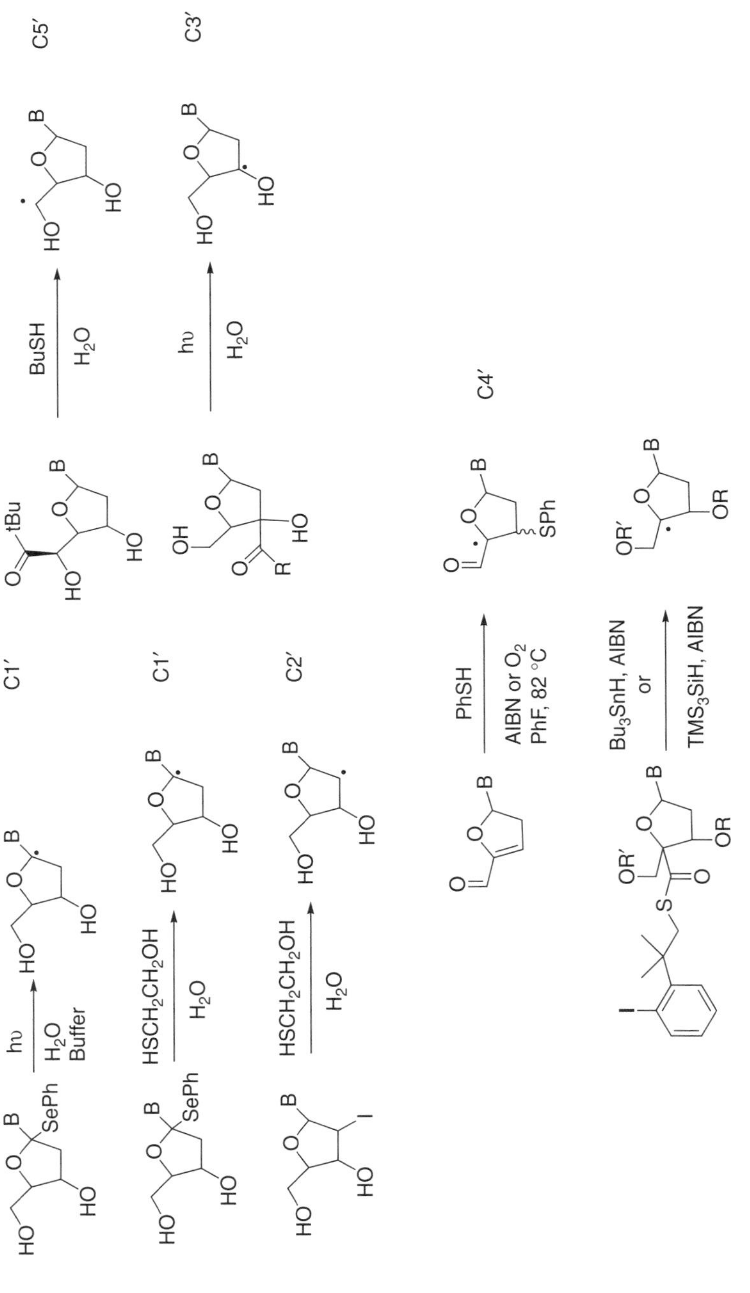

Scheme 10.1   Various methods of generation of C1′, C2′, C3′, C4′ and C5′ radicals from suitable radical precursors.

Aldehydes, ketones and esters

$R_2C=O \rightarrow R_2C-O^{\bullet}R \longrightarrow$

- N_2H_2, Hg^{2+}, $NaBH_4$
- Cathode, H^+
- R–H, photolysis
- $(COCl)_2$, N-hydroxypiperidine-2-thione, initiation
- Zn, TMSCl; Mg, TMSCl

$R_2C=O \rightarrow R_2C^{\bullet}-O^-R \longrightarrow$

- Na, Mg, Ti
- R_3N or HMPT, *hv*
- Cathode

10.2.3   Carbon–Sulfur and Carbon–Selenium Bonds

Alkyl sulfides, aryl sulfides, alkyl selenides and acyl selenides

$R-SC_6H_5$
  or $\longrightarrow R^{\bullet}$
$R-SeC_6H_5$

- Bu_3SnH, initiator
- H_3PO_2, initiator
- Anions, photolysis

10.2.4   Carbon–Nitrogen Bonds

Amines

$R-NH_2 \rightarrow R^{\bullet}$

- MeI, anion, photolysis or Na

Nitro compounds

$R-NO_2 \rightarrow R^{\bullet}$

- Bu_3SnH, AIBN
- TMS_3SiH, initiator
- Anion, *hv*

Diazonium salts

$$Ar-N_2^+X^- \longrightarrow Ar^\bullet$$

- Cu^+, Ti^{3+}
- Photolysis and thermolysis

10.2.5   Carboxylic acids

$$R-CO_2H \longrightarrow R^\bullet$$

- *N*-Hydroxypiperidine-2-thione, *hv*
- Mn^+, peroxide
- SO_2Cl_2, H_2O_2 or RO_2H, Δ
- Anode

10.2.6   Ketones

- H_2O_2, Fe^{2+}

10.2.7   Cyclopropanes

- Hg^{2+}, $NaBH_4$

10.2.8   Alkenes

- Hg^{2+}, $NaBH_4$
- B_2H_6, O_2; B_2H_6, Hg^{2+}, $NaBH_4$
- Peroxide, Mn^+

- Captodative alkenes
- Cyclisation of dienes, tandem cyclisation, combination of intramolecular and intermolecular reactions

10.2.9 Alkenes and Aromatic Compounds

$$\bigg\rangle\!\!=\!\!\bigg\langle \longrightarrow \bigg\rangle\overset{+}{C}\!\!-\!\!\overset{\bullet}{C}\bigg\langle \ \text{or} \ \bigg\rangle\overset{-}{C}\!\!-\!\!\overset{\bullet}{C}\bigg\langle$$

- Anode
- Cathode

10.2.10 Carbon–Hydrogen Bonds

$$R\!-\!H \longrightarrow R^{\bullet}$$

- Peroxides
- Ketone/hv or hv
- Mn^+
- R_2N^+HCl, Fe^{2+}
- Thermal

10.2.11 Carbanions

$$R^- \longrightarrow R^{\bullet}$$

- Anode
- I_2 or O_2

10.2.12 Carbon–Boron Bonds

$$R\!-\!B\!\big\langle \longrightarrow R^{\bullet}$$

- O_2
- Hg^{2+}, $NaBH_4$

10.2.13 Carbon–Mercury Bonds

$$R\!-\!HgX \longrightarrow R^{\bullet}$$

- NaBH$_4$ or Bu$_3$SnH
- Anions *hv* or heteroaromatic/*hv*.

10.2.14   Carbon–Cobalt Bonds

$$R-Co \longrightarrow R^\bullet$$

- Thermal

References

1. Y. Xu and H. Sugiyama, *Agnew. Chem. Int. Ed.*, 2006, **45**(9), 1354–1362.

Further Reading

C1′ radicals: T. Gimisis, G. Ialongo, M. Zamboni and C. Chatgilialoglu, *Tetrahedron Lett.*, 1995, **36**, 6781–6784; T. Gimisis, G. Ialongo and C. Chatgilialoglu, *Tetrahedron*, 1998, **54**, 573–592.

C2′ radicals: C. Chatgilialoglu, M. Duca, C. Ferreri, M. Guerra, M. Ioele, Q. G. Mulazzani, H. Strittmatter and B. Giese, *Chem. Eur. J.*, 2004, **10**, 1249–1255.

C3′ radicals: K. Steffi, A. Bryant-Friedrich and B. Giese, *J. Org. Chem.*, 1999, **64**, 1559–1564.

C4′ radicals: B. Giese, P. Erdmann, T. Schaefer and U. Schwitter, *Synthesis*, 1994, 1310–1312; B. Giese, A. Dussy, C. Elie, P. Erdmann and U. Schwitter, *Angew. Chem. Int. Ed.*, 1994, **33**, 1941–1944; B. Giese, A. Dussy, E. Meggers and M. Petretta, *J. Am. Chem. Soc.*, 1997, **119**, 11130–11142; D. Crich, Q. Yao, *Tetrahedron*, 1998, **54**, 305–318; M. L. Javacchia and P. C. Montavecchi, *Org. Biomol. Chem.*, 2006, **4**, 3754–3756

C5′ radicals: A. Manetto, D. Georganakis, L. Leondiadis, T. Gimisis, P. Mayer, T. Carell and C. Chatgilialoglu, *J. Org. Chem.*, 2007, **72**, 3659–3666.

Subject Index